NERVE

ALSO BY TAYLOR CLARK

Starbucked: A Double Tall Tale of Caffeine,
Commerce, and Culture

NERVE

POISE UNDER PRESSURE,
SERENITY UNDER STRESS, AND
THE BRAVE NEW SCIENCE OF
FEAR AND COOL

TAYLOR CLARK

LITTLE, BROWN AND COMPANY
New York   Boston   London

Little, Brown and Company
Hachette Book Group
237 Park Avenue, New York, NY 10017
www.hachettebookgroup.com

First Edition: March 2011

Little, Brown and Company is a division of Hachette Book Group, Inc.
The Little, Brown name and logo are trademarks of Hachette Book Group, Inc.

The publisher is not responsible for websites (or their content) that are not owned by the publisher.

Library of Congress Cataloging-in-Publication Data
Clark, Taylor.
    Nerve : poise under pressure, serenity under stress, and the brave new science of fear and cool / Taylor Clark. — 1st ed.
        p. cm.
    Includes bibliographical references and index.
    ISBN 978-0-316-04289-5
    1. Fear. 2. Anxiety. 3. Calmness. I. Title.
    BF575.F2C525 2011
    152.4'6 — dc22   2010038835

10 9 8 7 6 5 4 3 2 1

RRD-IN

Printed in the United States of America

Anything scares me, anything scares anyone but really
after all considering how dangerous everything is
nothing is really very frightening.
— GERTRUDE STEIN

Contents

NERVE

The Closest Call

On the evening of October 27, 1962, as his debilitated submarine lay pinned deep below the Sargasso Sea by American destroyers, their thunderous depth charges shredding his nerves blast by earsplitting blast, Captain Valentin Savitsky of the Soviet Northern Fleet finally lost his cool. And when Valentin Savitsky lost his cool, the world itself nearly went to hell.

From the very moment Savitsky and his comrades first set sail, their mission had appeared to be doomed. The crew's ordeal began three weeks earlier, when their vessel, *B-59*, departed from its frigid home port of Sayda Bay alongside three other *Foxtrot*-class submarines for what everyone aboard assumed would be a series of dull training exercises off the Siberian coast. The Soviet commanders, however, had been strangely vague about their orders; they gave each captain a set of secret instructions that were to be opened only at sea, and simply told them to turn left after they exited the bay. In truth, if these commanders had issued their cloak-and-dagger orders in person, the submarine crews might have gasped in disbelief: the four outdated diesel-electric subs, which had been designed only for operations in cold northern waters, were to race five thousand miles southwest—avoiding detection all the while—and establish a spearhead for a future naval base near Havana, Cuba. With Soviet ballistic missile launch sites already sprouting up all over the Cuban mainland, *B-59* was about to become a prop in a terrifying game of nuclear chicken between two stubborn superpowers: the Cuban Missile Crisis.

No one expected it to be a pleasure cruise, but even still, the *Foxtrot* quartet's voyage southwest was exceptionally brutal. In order to hit their maximum speed and reach Cuba quickly, the Soviet boats (as submariners refer to their vessels) needed to sail on the water's surface—right through Hurricane Daisy, which rocked the Atlantic Ocean with violent fifty-foot waves that would suddenly topple the vessels onto their sides. The tempest played havoc with the subs' engineering systems, slamming air intake valves shut with enough force to create boat-wide vacuums that sucked the fillings out of the sailors' teeth. Meanwhile, NATO tracking planes hunted them at every turn. When at last the four Soviet vessels reached the calmer tropical seas undetected, the crews thought the worst had passed. And that's when their submarines began to fall apart.

For the *Foxtrot* boats, the equatorial water may as well have been poison. Even five hundred feet below the surface, water temperatures hovered above 80 degrees Fahrenheit, and the tropical oceans contained more salt than the cold-water subs could handle; working together, the salinity and the heat gradually crippled the Soviet boats. Excess salt blocked coolant and air circulation systems, shredded rubber sealings, and incapacitated engines. The temperature in some sections of the subs soared above 120 degrees, giving the crewmen horrendous rashes and sapping their strength. "My head is bursting from the stuffy air," wrote one sailor in his diary. "Today three sailors fainted from overheating again.... Those who are free from their shifts are sitting immobile, staring at one spot."

Despite the punishing conditions, the four Soviet submarines came tantalizingly close to reaching Cuba, when a puzzling radio transmission from Moscow directed them to *turn around* and go patrol the Sargasso Sea, due east of Florida. To the crews, this plan seemed pointless and suicidal; all of the radio reports they received from home were about mundane things like crop yields from the fall harvest, and thus they had no idea that an international crisis was afoot. Ironically, they found out about the nuclear standoff only after they intercepted transmissions of the "Voice of America," the United States government radio service—broadcasts that specifically outlined President John F. Kennedy's determination to prevent Soviet submarines from operating in American coastal waters. As the *Foxtrot* captains soon learned, Kennedy had commanded the U.S. Navy to spare no effort in tracking down their four rickety boats.

On board *B-59*, things swiftly went from bad to worse. As the sub glided through the deeps of the Sargasso, its surface radar showed a dozen American blips chasing them: destroyers, aircraft carriers, Neptune patrol planes, helicopters, all dropping blinking sonobuoys into the sea to ferret out their position. Whenever *B-59* surfaced for a moment to ventilate the boat and recharge its drained batteries, an American sighting inevitably forced them back down. The oppressive heat only intensified, and the crew wheezed and panted from the noxious, carbon-dioxide-rich air. The Soviet sailors, one man reported, "were falling like dominoes." On October 27, after days of cat-and-mouse, a U.S. plane finally spotted *B-59* breaching beneath an overcast sky. The American fleet immediately surrounded the beleaguered submarine as it dove below in a vain attempt to escape. Now, Savitsky and his crew found themselves not just trapped; to their shock, the Americans seemed to be *attacking* them. Days before, the Pentagon had sent a message to the Soviets, warning them that American ships would use harmless practice depth charges to signal the Soviet submarines to come to the surface. But with Soviet interpersonal communication skills being, shall we say, in need of polishing, the Soviet brass never relayed the message to the sub commanders. Instead, Savitsky and his crew believed they were taking part in the beginning of World War III.

So far, this tale probably sounds like any other tense encounter that occurred during the thirteen days of the Cuban Missile Crisis, and for decades, historians assumed that these American destroyers were merely harassing a few overmatched, wretched little submarines. Because the Soviets had sent such antiquated vessels, the Americans assumed they packed an antiquated punch—standard torpedoes, at best. But what the Americans didn't know about these four submarines could have destroyed the world: each *Foxtrot* sub carried one torpedo tipped with a fifteen-kiloton nuclear warhead, boasting the same explosive power that leveled Hiroshima. If fired, one of these torpedoes could annihilate everything within a ten-mile radius. What's more, the Soviet commanders had given the submarine captains vague, even ominous instructions for using the warheads. In a briefing before their departure, one official told the captains to fire the nuclear torpedoes only if the Americans blew a hole in their hull or if they received direct orders from Moscow, but added, "I suggest to you, commanders, that you use the nuclear weapons first, and then you will figure out what to do after that." Another admiral went one step further. "If they slap you on the left cheek," he said, "do not let them slap you on the right one." By harassing *B-59*, the Americans thought they were needling a relative weakling. In reality, they were poking a dragon with a twig.

With this nuclear tidbit in mind, let's review the psychological situation Savitsky faced under the dim lights of his cramped, muggy command post. Fear? Check: Savitsky and his men had received no information from Moscow, and for all they knew, they were about to plummet to the ocean floor in the opening salvo of the next world war. Stress? Check: the men were exhausted and unwell after three horrendously punishing weeks, and the exploding U.S. depth charges sounded, in one crewman's words, "like you were sitting in a metal barrel, which somebody is constantly blasting with a sledgehammer." Pressure? Check: for Savitsky, to fail his motherland against its sworn enemy would be a fate worse than death. And thus, Captain Valentin Savitsky, stalwart commander of the Soviet sub *B-59*, finally lost his cool, cracking under the fear, stress, and pressure of a situation no man should ever have to confront. Savitsky summoned his staff to the command post and, banging the table with his fists, roared in fear and anger about the Americans torturing them from above. "Maybe the war has already started up there, while we are doing somersaults down here!" he cried. "We're going to blast them now! We will die, but we will sink them all—we will not disgrace our navy!"

But then something incredible happened. Into the midst of this terrifying scene, with Savitsky yelling emotionally and the crewmen convinced they were about to perish in a nuclear blast, walked a man named Vasily Arkhipov. A soft-spoken, slightly older officer, Arkhipov was the chief of staff of the submarine flotilla, traveling on *B-59* as a guest. For his gregarious nature and his ability to keep calm under pressure, Arkhipov had won the admiration and respect of all who served with him. Ignoring the panic in the command post and the depth charges bursting around the ship, Arkhipov took Savitsky aside and sat with him for a few moments, speaking to him in soothing, simple words. His cool manner calmed Savitsky down, and they soon agreed that it would be best to come up and talk to the Americans before doing anything rash—even though being forced to the surface by the enemy is the highest humiliation a submarine commander can suffer. Gritting his teeth, Savitsky gave the order to ascend and greet their tormentors.

The Americans didn't make it easy on him. Savitsky's submarine emerged with an enormous gush among four destroyers, spotlights from navy helicopters beaming down upon them. *B-59*'s haggard crew climbed up and gasped in the cool night air. After taunting them with explosives for four hours, the commander of the USS *Lowry* now archly asked Savitsky if he needed assistance. Completely oblivious to their near vaporization moments before, the Americans sent a jazz band onto the deck to play "Yankee Doodle," as U.S. sailors danced and tossed packs of cigarettes and cans of Coke down to the sub, most of which landed in the water. (Savitsky, admonishing his crew to "behave with dignity," banished one man back inside for tapping his foot to the music.) Within a couple of days, *B-59* dove and eluded the Americans once again, and the four subs returned home not to a hero's welcome but in disgrace. Gaunt and half dead from their ordeal, the *Foxtrot* commanders faced interrogation after interrogation by party officials, who gravely informed them that they had shamed the Soviet Union by not demonstrating her nuclear power. Said *B-4*'s Captain Ryurik Ketov of this "hell," "We would rather go through another Cuban campaign than to go through again what we went through when we got back." Their "failed" mission disappeared from official Soviet records, and Savitsky and the other officers were shuffled off into permanent obscurity. Vasily Arkhipov died in 1999, three years before the public first learned that the *Foxtrot*s carried nuclear weapons. The only person he had told about his heroism was his wife.

The world had many opportunities to blow itself up during the Cuban Missile Crisis, but never before had one man, in the most frightening and stressful conditions imaginable, made such an incredible impact by keeping cool under fire. Without Arkhipov's intervention, twentieth-century history might have unfolded far differently. Captain Savitsky didn't give the final order to arm the torpedo, and his finger was never quite perched over the glowing red button of nuclear annihilation, but what would have happened if the situation on *B-59* had continued to spiral out of control? Svetlana Savranskaya, a Russian-born scholar of the crisis who has interviewed all of the living *Foxtrot* officers, sketches a grim portrait of the likely sequence of events. "Just imagine, the Americans didn't know these subs carried nuclear weapons, so the shock factor would have been immense," she told me. "If a nuclear torpedo had been used from a submarine, it would have provoked such a panic that the Americans would have immediately launched a ground invasion of Cuba. And then, of course, the tactical nuclear weapons stationed there would have been used against the invading force, and in the next turn of the spiral the Americans would have felt they had to use nuclear weapons themselves." Or, in the words of Thomas Blanton, director of the National Security Archive at George Washington University, "The lesson from this is that a guy named Vasily Arkhipov saved the world."

But let's be a bit more precise. What saved the world was Arkhipov's extraordinary ability to maintain his poise in a colossally frightening and tense situation—one that made even the iron-willed Valentin Savitsky buckle under the strain. So how did he do it? How does *anyone* do it? What is it that enables a person like Vasily Arkhipov to keep his head and do the right thing despite the storm raging within?

This is a book about how people deal with fear, anxiety, stress, and pressure in all of their forms. Obviously, few of us will ever encounter all of these forces at once in the psychological super-nova that Savitsky and Arkhipov endured on *B-59*, yet we don't exactly require the prospect of global thermonuclear catastrophe to get us to twist ourselves into impressive emotional knots. Nerves make us falter through job interviews, choke on tests in school, and lose it in traffic jams. An upcoming presentation for a dozen people at the office can rob us of sleep and peace of mind for weeks, as we fret in circles over everything that could go wrong. Even something as innocuous as a blind date can send us into fits of worry.* In all walks of life, fear and stress loom on the horizon: they freeze cops in tight situations, paralyze concert performers on stage, and make skydivers' brains lock up so much that they can forget to pull their parachutes. No one is immune. The celebrated British actor Stephen Fry once experienced such sudden and intense stage fright that he fled not only the stage but *the country* in the middle of a theatrical production; Mike Vanderjagt, the most accurate placekicker in National Football League history, missed a single clutch field goal try and never recovered his accuracy. In frightful and nerve-racking times, it sometimes seems that our minds transform into anxiety-fueled arch-villains, determined to foil our deepest hopes.

When you think about it, it's one of the great ironies of our time: we now inhabit a modernized, industrialized, high-tech world that presents us with fewer and fewer legitimate threats to our survival, yet we appear to find more and more things to be anxious about with each passing year. Unlike our pelt-wearing prehistoric ancestors, our survival is almost never jeopardized in daily life. When was the last time you felt in danger of being attacked by a lion, for example, or of starving to death? Between our sustenance-

---

*And here, I realize many readers would protest that there is no such thing as an innocuous blind date.

packed superstores, our state-of-the-art hospitals, our quadruple-crash-tested cars, our historically low crime rates, and our squadrons of consumer-protection watchdogs, Americans are safer and more secure today than at any other point in human history.

But just try telling that to our brains, because they seem to believe that precisely the opposite is true. At the turn of the millennium, as the nation stood atop an unprecedented summit of peace and prosperity, anxiety surged past depression as the most prominent mental health issue in the United States. America now ranks as the most anxious nation on the planet, with more than 18 percent of adults suffering from a full-blown anxiety disorder in any given year, according to the National Institute of Mental Health. (On the other hand, in Mexico—a place where one assumes there's plenty to fret about—only 6.6 percent of adults have *ever* met the criteria for significant anxiety issues.) Stress-related ailments cost the United States an estimated $300 billion per year in medical bills and lost productivity, and our usage of sedative drugs has shot off the charts: between 1997 and 2004, Americans more than doubled their yearly spending on antianxiety medications like Xanax and Valium, from $900 million to $2.1 billion. And as the psychologist and anxiety specialist Robert Leahy has pointed out, the seeds of modern worry get planted early. "The average high school kid today has the same level of anxiety as the average psychiatric patient in the early 1950s," he writes. Security and modernity haven't brought us calm; they've somehow put us out of touch with how to handle our fears.

It wasn't supposed to be like this. After all, fear is truly our most essential emotion, a finely tuned protective gift from Mother Nature. Think of fear as the body's onboard security system: when it detects a threat—say, a snarling, hungry tiger—it instantly sends the body into a state of high alert, and before we even comprehend what's going on, we've already leapt to the safety of a fortified Range Rover. In this context, fear is our best friend; it makes all of the major decisions for us, keeps the personage as free of tiger claws as possible, and then dissipates once the threat has subsided. The problem, though, is that the fear system wasn't designed for modern life. When our world's comparatively tame hazards and worries present themselves—*Does my boss hate me? Will my flight be delayed? Did the waiter hear me say "No chipotle mayo"?*—our bodies still react as though we're staring down that ravenous tiger; we're working with the same neural technology as our hunter-gatherer ancestors. This technology wasn't engineered to help us do calculus, perform a flawless piano sonata, or throw a pinpoint fastball under pressure, yet that's what we often ask of ourselves.

Cool-headed people who navigate gracefully through unimaginably fearsome situations often become our greatest heroes, but we generally have little idea *why* they display such poise. We flock to movie theaters to watch tales of soldiers who keep their wits about them despite bullets whizzing by, of unflappable trauma doctors whose hands are statue-steady during lifesaving surgery, and of international spies who yawn as they foil evil plots for global domination. We marvel at the mental toughness of Kobe Bryant as he sinks yet another three-pointer with the game on the line—but seriously, what on earth *is* "mental toughness"? A state of mind? A God-given personality trait? A skill anyone can learn? A benefit of purchasing Nike footwear? We seldom get solid answers, just a slew of clichés. One thing we know for sure is that these exceptional people aren't *immune* to fear. For example, when the astronaut Neil Armstrong was maneuvering his fuel-starved lunar module through treacherous fields of boulders to find a safe landing spot on the surface of the moon, the *Eagle*'s instruments showed that his heart was thumping out of his chest, yet his voice was so calm that it was as though he were piloting a commuter flight to Omaha. This kind of composure seems almost inhuman; no wonder we speak of those like Armstrong as having "ice water in their veins." We're reverent because their poise and their tolerance for terror absolutely mystify us.

When it comes to dealing with our own fears, though...well, let's just say that many of us have a true gift for digging ourselves even deeper into the hole. Our widespread troubles with anxiety and stress have escalated in large part because we tend to address them in ways that actually make things worse. We struggle to control our emotions, and we avoid the situations that make us anxious—two strategies that are destined to fail. We try to dull our nerves with alcohol or other substances. We give each other useless nuggets of advice, like "Just try to relax." ("Ah, of course! Why didn't I think of that?") We worry in endless loops about the things that scare us, without realizing that this fretting never yields anything productive. We pay lip service to the idea of facing up to our fears, but in practice we steer clear of doing any such thing. In short, we've taken our most indispensable and useful emotion and turned it into our foe.

Before we go any further, I want to be perfectly clear about one thing. During the research process for this book, as I'd be interviewing an expert or telling a friend about my latest findings, people frequently asked me the following question, sometimes delicately and sometimes with obvious skepticism: "What, exactly, makes you qualified to write a book telling people how they should deal with their fears?" To this, I generally replied, "I...wait, what? You're breaking up," and then quickly hung up the phone. (This tactic didn't work quite as well in person.) Because, to be honest, I am hardly the cool-headed master of fear. I'm not a psychologist, and I'm not a guru with a seven-step plan to help you End Worry Today! or Unleash Your Fearless Warrior Spirit. I have never scaled a sheer cliff face without safety ropes, single-handedly destroyed an enemy tank, or thrown a last-second touchdown pass. I am, in fact, a fairly neurotic guy with more than my fair share of irrational, deep-seated worries and anxieties—hell, I'm even nervous as I write these words. So, to answer the question, I am *not at all* qualified to tell anyone how to expertly manage their fears.

Thanks for picking this up! It's been fun! Enjoy browsing the rest of the bookstore!

Actually, there's a good reason why a curious journalist like myself—one with plenty of anxious firsthand experience and a strong desire to uncover concrete truths in an often hazy subject—might be a good guide through the trembling landscape of fear. A few years ago, for an assortment of reasons I won't bore you with, I went through a long stretch of persistent anxiety. My best efforts to help myself proved maddeningly useless; the harder I tried to figure my fears out or drive them away, the deeper they seemed to entrench themselves. I just didn't understand my own mind. When I turned to the bookshelves for guidance, I never quite found what I was looking for. I paged through vague self-help books that offered advice without scientific backup or sold a psychological bill of goods. Other books actually seemed designed to torture the anxious; I remember seeing one aged volume with the ominous title *YOU MUST RELAX* and thinking to myself, or else…*die?* I found dry, clinical books written by psychologists who seemed never to have experienced real anxiety; technical neuroscience tomes that felt divorced from the real world; and inspirational books stuffed with syrupy clichés. To be sure, there are a handful of wonderful, life-changing guides to fear and stress out there, but I was looking for something else—something broader. Watching a pro baseball player at the plate or a politician speaking before a huge crowd, I'd wonder, *How is this person not melting under the pressure right now?* Hearing of a soldier's heroic deeds in battle or a pilot's miraculous performance aboard a failing plane, I'd ask, *What makes a person capable of keeping cool and doing their duty in a terrifying situation like that?* I wanted a fuller picture of fear.

Fortunately—and not a moment too soon—a flood of cutting-edge research from psychologists, neuroscientists, and scholars from all disciplines is now coming together to show us what fear and stress really are, how they work in our brains, and why so much of what we thought we knew about dealing with them was dead wrong. Picking a painstaking trail through the labyrinth of the brain, a neuroscientist from the bayou traces our mind's fear center to two tiny clusters of neurons, uncovering the subconscious roots of fear. Using a simple thought experiment, a Harvard psychologist discerns why our efforts to control our minds backfire, and why a directive like "just relax" can actually make us more anxious. Employing one minor verbal suggestion, a group of Stanford researchers find they can make young test-takers' scores plummet in a spiral of worry—or hoist them right back up. Across the nation, intrepid scientists are discovering why athletes choke under pressure, how the human mind transforms in an emergency, why unflappable experts make good decisions under stress, and how fear can warp our ability to think.

Eighty years ago, President Franklin Roosevelt told a nervous nation, "The only thing we have to fear is fear itself," but it turns out that this isn't quite right. Our problem is almost never "fear itself" but the way we relate to that fear—by avoiding, withdrawing, seeking control, worrying, or falling victim to the mistaken belief that things will be okay only after we've annihilated all anxiety. Fear can be a very good thing: it helps us survive, gives meaning to our achievements, facilitates our performance, and makes us feel alive. Yes, fear can be uncomfortable and bewildering, and it can even thwart our most dearly held goals—but it doesn't have to be so. As I've confessed, I'm still learning to live in harmony with my own anxieties, and it's a relationship that will take many years to develop. You'll never hear me say that shaking hands with fear is an easy and painless process. Yet if I've ascertained anything from researching and writing this book, it is this: Fear is not our enemy. We don't need to get rid of fear or push it away. We need to learn how to be afraid.

This is why I've chosen to title this book *Nerve*, a word with two meanings that seem, on the surface, to directly conflict with each other. Having "a case of nerves" is a common synonym for fear itself, yet "showing nerve" signifies moral courage. In *Nerve*, I hope to show that these definitions are really two sides of the same coin. Fear is an essential part of courage and poise, and the more we learn to work with it—showing nerve with our nerves, as it were—the further our horizons expand. No one can become as cool-headed as Vasily Arkhipov on *B-59* overnight, yet each and every one of us can make significant, even transformative improvements in how we relate to our anxiety and stress. We might not end up saving the world, but we may very well end up saving ourselves.

PART ONE

The Nervous Trinity

Fear

Anxiety

Stress

CHAPTER ONE

Your Second Brain: Exploring the New Science of Fear

The first thing Scott Raderstorf remembers about the most terrifying day of his life was the faint tremor of his bed beneath him, a vibration so slight that he assumed it was a rat gnawing at his bedpost. In the groggy early-morning light, this seemed the most logical explanation; Raderstorf and his still dozing family were lodging for the week in an open-air thatched bungalow on a crescent of unspoiled Thai beach, and occasional rodent incursions were to be expected. When he glanced at the bedside table, though, Raderstorf noticed a flutter in his glass of water as well—so unless this was a particularly brawny rat, the trembling was coming from elsewhere. Maybe someone was out dynamite fishing, he reflected. Raderstorf blearily considered arising for a dawn-cracking kayak trip but thought better of it and drifted back to sleep. After all, he was on vacation.

More specifically, the Raderstorfs were on vacation from their vacation. At age forty-two, Scott Raderstorf was living out a dream that he and his wife, Joellen, had schemed over for years: they were taking their three young sons on a round-the-world adventure. Having recently sold off a software company, the couple seized on their newfound freedom and charted a journey that would carry them away from their Boulder, Colorado, home for six months — Dubai, Melbourne, Istanbul, Beijing, Johannesburg, and more. They set off in November 2004 with a single backpack each, blitzing first through Japan, then China, all of it more or less improvised. "The only thing we had were the plane tickets," explained Raderstorf. "Other than that, we just showed up and figured it out." The one pre-planned reservation that the Raderstorfs allowed themselves was for a Christmas sojourn at the Golden Buddha Beach Resort on pristine Koh Phra Thong island, three hundred miles southwest of Bangkok. Arriving at the retreat via longboat amid an idyllic Christmas Eve sunset, the family felt they had reached heaven itself.

By the time the Raderstorf clan ventured out for breakfast on their third day at the Golden Buddha, Scott had forgotten all about the morning's mysterious rumblings. It was another flawless day in paradise, and after making himself a press pot of coffee, Scott took his laptop to the Internet shack while the two youngest, Quin and Max, headed for the beach to make castles in the fluffy beige sand. On a call to the friends who had told them about the resort, the trouble began. "I called to say 'Merry Christmas and thanks for the recommendation,'" Raderstorf told me. "I'd literally just gotten that much out when I heard this huge boom down on the beach, like a jet plane had crash-landed. So I ran out, and my wife ran up from the breakfast area, and this big, broad beach was totally wet." The sea had been as placid as bathwater before, yet a massive wave had suddenly flooded forty feet up the beach and swiftly drained away. The quaking of a few hours before flashed forebodingly through Scott's mind. In the water, Scott saw people frantically swimming toward the shore as a powerful current swept them out to sea. And where Quin and Max's sandcastle had been a moment before there was now only a waterlogged mess.

This was Scott Raderstorf's warm-up fright of the day. With their oldest, Ben, in tow, Scott and Joellen tore down the beach, yelling their sons' names. The boys weren't there. The family's bungalow was empty. Their parental alarm amplifying, Scott and Joellen were scanning the sea with darting eyes when the boys appeared in the entrance of the cottage two doors down, guilty looks on their faces. By an incredible coincidence, the Raderstorfs had run into two old friends at the Golden Buddha, Roger Hodgson and Carrie Sengelman, and their kids' irresistible Game Boy—which Quin and Max had been sternly forbidden to play—likely saved the Raderstorf children's lives.

It was while the two families were exchanging relieved hugs that Carrie noticed something peculiar: the ocean was *disappearing*. The waterline slid farther away by the second, first a hundred yards distant, then two hundred, leaving schools of fish flopping on a vast, smooth expanse of sand. The sight mesmerized Scott and Carrie—but not Joellen, who immediately started shepherding the kids away from the beach. With the water now a quarter mile away, Roger hurried into his family's bungalow to secure his laptop, but Scott wandered forward to watch the entrancing spectacle. Like several others on the beach, he saw little to be concerned about. "It didn't *look* menacing," he recalled. "It was just novel. I knew the water would come back, but I thought at worst the resort might get a little swamped."

When the wave of returning water first loomed on the horizon, it appeared small and almost tranquil, a burbling, seemingly harmless white tide. As the seconds ticked by, however, the water grew strangely loud, sending out a rumble that struck Raderstorf as far too deep for such a tiny wave—and anyway, shouldn't it have arrived by now? The truth didn't click in his mind until he saw the wave pass a large rock at sea, and the water that had once appeared so sedate enveloped it violently and completely. In an instant, his mind recomputed the scale of what he was looking at: it wasn't a sluggish little wash of water; it was a thirty-foot-tall mountain of surf moving with crushing speed. The vibration he had felt that morning wasn't a rat; it was the radiating tremor of the second-largest earthquake ever recorded, issuing from deep beneath the Indian Ocean. And the wave it triggered wasn't a wave at all. It was a tsunami—and he was standing directly in its path.

Right then, Raderstorf's brain grasped that he was probably about to die. A biochemical switch flipped. Within milliseconds, his body flooded with the strongest feeling of fear he had ever experienced, a physiological torrent that transformed him at once into a primal creature with a single overpowering urge: *survive*.

And...let's freeze the story right here.

We'll return to Raderstorf's bout with the Boxing Day tsunami in a moment, but first, let's ask a question: What in the world just happened to him? How, literally in the blink of an eye, did he swing from a state of mesmerized awe to one of abject terror, with adrenaline coursing through his every limb? What, in other words, *is* this fear he felt, and how does it work? We generally don't think in such an analytical way about our emotions, as nebulous and unpredictable as they often seem to us, but in Scott Raderstorf's case it's no idle question. Without a swift and powerful fear response, flipped at just the right moment and precision-engineered to produce the exact sequence of physiological events that soon followed, he would have died that day. Raderstorf's case may be extreme, but his story also offers us the clearest possible illustration of why our fear is a very good thing.

In this chapter, we will explore the split second when fear bursts into being within our bodies and brains, a surprisingly rich and complex sliver of time that neuroscientists have only recently demystified. Thirty years ago, the scientific establishment believed emotions to be too troublesome to study with any rigor, yet thanks to technological advances and a few innovative minds, we now understand how fear works, where it lives in the brain, and why it behaves in such puzzling ways. What, for example, would make an arachnophobic terrified of a toy tarantula, even though it posed no threat to his safety? Why doesn't reason make fear immediately dissipate? How is it that we can react to danger before we've actually figured out what's going on? It all happens because our fear mechanism has a mind of its own—in fact, it *is* a mind of its own. Before we can comprehend the secret ingredients of clutch, cool, and grace under fire, we first need to understand fear itself.

The Three *F*s

Our systematic knowledge about the circuitry of fear might be brand-new, but long before neuroscience ever beamed a spotlight on the emotion's inner workings, we'd already pieced together a tidy explanation for its purpose in human life. Of course, philosophers have pondered the nature of fright and terror since the days of Plato; Spartan warriors even built temples in honor of fear. Our first truly useful insights about the function of fear, however, trace back to the dawn of evolutionary thought.

Working from the assumption that all of our emotions must have arisen to fulfill some survival-ensuring purpose, Charles Darwin (among others) loved to muse about why certain situations make us feel the way we do. In his 1872 book *The Expression of the Emotions in Man and Animals*, Darwin wrote of one particularly revealing experiment he attempted at the Zoological Gardens in London. Pressing his face up close to the thick glass wall of a puff adder cage, Darwin set his concentration toward one goal: if the venomous (and highly annoyed) snake tried to strike, he would not allow himself to flinch. Time after time, no matter how strong his determination, he got the same result. "As soon as the blow was struck," he wrote, "my resolution went for nothing, and I jumped a yard or two backwards with astonishing rapidity." Darwin marveled at the sheer strength of his fear response, against which his "will and reason were powerless." Clearly, he was dealing with a potent, deeply ingrained emotional reaction—maybe the most important one of them all.

To better demonstrate the precise evolutionary function of fear, let's travel back in time, oh, a million years or so to pay a visit to one of our Stone Age ancestors. We'll call him Thag. Feeling pent up in his damp and poorly furnished cave one day, Thag decides to take a restorative stroll through a nearby glade. Mere minutes into his walk, though, an unforeseen danger streaks through the underbrush: a mastodon charging straight for him, angry puffs of steam billowing from its trunk. In this dire predicament, with not even a spear at his disposal, Thag can do two things—think or act. Suppose Thag chooses Option A and tries to reason his way out of the situation. His thinking might look something like this: *Ah yes, the mighty mastodon. Majestic creature. He appears agitated with me. Now, then: how to survive? How . . . to . . . sur—* And right about here, Thag would be crushed underfoot. Put simply, thinking in this situation is maladaptive; it wastes precious time, stands in the way of useful action, and puts the Thag bloodline in jeopardy. To maximize his chances of staying alive, Thag needs to have some kind of self-activating defense system in place that can instantly supercharge his body and launch it into a protective action like sprinting away—preferably before he even has an *opportunity* to ponder other options. He needs fear.

Fear is nature's little way of telling us the following: Sorry, but you're not to be trusted with your own survival. It's a blunt physiological tool designed to automatically supersede every other bodily function and ensure our continued existence *right now.* "Fear evolved as a protective mechanism focused on antipredator behavior, and it's one of the most powerful forms of selection pressure," explained Michael Fanselow, a psychology professor at the University of California, Los Angeles, who is one of the nation's foremost experts on the science of fear. "So here's the example I typically give to illustrate that. We've all missed a mating opportunity, but we'll have another one. If we miss a feeding opportunity, with someone like me that's probably a good thing—it's not going to hurt us. But if we miss an opportunity to defend ourselves, that's it. And we don't reproduce anymore." Because of this premium on snap-quick, decisive action, evolution programmed our fear system to react first and ask questions later. "You want to be conservative here, right?" Fanselow continued. "If something has the potential to do you harm, you want to rapidly and completely react to it."

In 1915, the Harvard University physiologist Walter Cannon coined a catchy term that we still use to describe this automatic fear system: the fight-or-flight response. Cannon's fundamental insight was that all animals, from brainy humans down to tiny field mice, meet a sudden threat with the same basic biological reaction—a state of impulsive nervous excitement that prepares them either to scamper away at warp speed or to battle tooth and claw. Adrenaline surges through the body, charging the muscles with energy for emergency action. Surface-level blood vessels constrict and leave the skin pale and slightly numb, providing a temporary layer of armor that is less likely to bleed.* The pupils dilate for

---

*This is why gunshot victims often don't even notice their injury at first: it's only when their blood vessels open again after the initial shock passes that the bleeding begins.

heightened vision. Nonessential processes like digestion cease—
and worse, the body sometimes decides to jettison all extra weight,
which accounts for the unfortunate sudden loss of bowel and bladder control that can strike when we're frightened. Breathing and
heart rate both speed up, funneling more oxygen into the muscles.
In short, you're ready to rock and/or roll. The brain decides
whether to fight or flee subconsciously and instantaneously, and
threatened animals often swing back and forth between the two.
Confronted by a hostile fox, a squirrel might first sprint away in
terror, then lash out in a reckless last-ditch attack after it's been
cornered.

As any fear researcher could tell you, however, Cannon missed
one important detail. There are really *three F*s to the fear response:
fight, flight, or *freeze*. Imagine that you're walking alone down a
dark street, and you're startled by a suspicious rustling off in the
bushes. What's the first thing you'll do? Without thinking, you
will stop dead in your tracks, your senses suddenly razor sharp.
Your eyes will fly wide open to take in as much visual information
as possible, and you'll quickly draw in and hold your breath, rendering yourself completely silent. Your mouth will drop open, in
case you need to scream for help. In a flash, you've become an
alarmed-looking human statue.

Based on what we've discussed so far, this freezing response
probably seems somewhat wrongheaded; wouldn't it be better to
get a head start and dash away from the potential threat immediately? Actually, no. In the wild, many predators react to movement, and if you abruptly go rigid there's always a chance that the
tiger you just spotted won't notice you. Think of freezing as a
state of defensive preparation. The body gets the same jolt of
adrenaline that readies it for fighting or fleeing, but the brain has
calculated that at least for the moment, your best odds of survival

come with no action at all.* Sometimes this gamble backfires, the most obvious example being deer that freeze in the headlights of an oncoming Mack truck, their brains programmed to hope this weird, shiny predator will fail to see them. We often dismiss the petrified deer's reaction as primitive or brainless, and then, of course, we do the very same thing when prodded to speak before an audience—most likely one that contains zero predators.

In an especially menacing situation, some animals stretch the freezing response even further and instinctively play dead, on the theory that predators have an innate revulsion to eating things that are already deceased (as an evolutionary safeguard against food poisoning). The North American hognose snake, for instance, takes the play-dead reaction to such comical heights that it resembles a standup comedy routine more than a survival behavior. If you startle a hognose as it winds through the grass, it will hiss and lash out just like any other snake—at first. This is mostly for show; the snake's bite packs no venom. So, when its tough-guy act fails to intimidate, the hognose opts for Plan B. Like a bad actor in a community theater production of *Hamlet*, the snake suddenly lurches as if struck by a mortal blow. Apparently afflicted with incredible pain, it writhes and twists dramatically before finally flopping on its back, its tongue hanging out the side of its gaping mouth: dead, from some invisible poison. It emits a rancid musk to complete the ruse, and sometimes drops of blood will even appear in its mouth. Turn it right side up, and the snake will instantly flip itself over again, as if to say, "No, seriously—I'm totally dead."

---

*Freezing can also happen when the brain gets overloaded by an unforeseen emergency situation and doesn't know what to do, a phenomenon we'll encounter in Chapter Seven.

Terrified humans feign death too, and not always by choice. Biologists call this kind of response tonic immobility, and we resort to it only when we're literally in the jaws of a predator. The celebrated explorer David Livingstone learned about this first-hand in 1844, when he attempted to help some African villagers fend off a lion and promptly found himself between the creature's teeth, being shaken "as a terrier does a rat." Almost immediately, Livingstone felt himself go limp and numb, later reporting that he remained fully aware through the entire experience; the ordeal, he wrote, seemed to occur through "a sort of dreaminess in which there was no sense of pain nor feeling of terror." The lion soon dropped him to pursue more active prey, and thus Livingstone's involuntary stupor may well have saved him. Likewise, many survivors of the 2007 Virginia Tech massacre—in which an unhinged gunman shot his way through several classrooms, killing thirty-two trapped students and professors before turning his gun on himself—recalled afterward that they fell to the floor and entered a catatonic trance when the assailant opened fire in their classroom, even though they hadn't been hit themselves. This too was adaptive, but it's also rare. Generally, freezing in humans is just a prelude to the two major survival actions of fighting or fleeing.

For Scott Raderstorf, caught on an exposed patch of sand as the tidal wave thundered in, the obvious choice was flight—and that's what he did, faster than he ever had in his life. Mere milliseconds after his brain comprehended the shocking scope of the oncoming torrent, Raderstorf's fear system shot him into action. Without thinking, he pivoted around and entered a dead sprint. "RUN!" he bellowed to Carrie, the family friend who also remained on the beach. "I remember the total panic," he told me of this moment. "I was just supercharged with adrenaline, more hopped up than I'd ever felt before." What he saw as he turned to run only deepened his terror: the nearest sturdy structure stood a hundred yards away, across a stretch of open beach. Scott spotted Joellen and the kids climbing a staircase to the top of the house—one of the few on the island with a brick foundation. He figured he had maybe fifteen seconds to get there before the wave crashed over the resort.

The loose sand made running nearly impossible, but Raderstorf's legs were set to turbo.* As Scott flew over the beach, he heard Carrie scream "HELP!" behind him—yet when he looked over his shoulder, he saw only a rumbling white wall of seawater. An instant after he passed the first row of thatched cottages, they exploded under the wave's force. "The first houses got hit by the front of the wave, and you could hear this huge boom as it blew these structures apart, turning them into toothpicks," Raderstorf recalled. Palm trees snapped in half all around him. By the time he reached the house, he could feel the spray from the tsunami on his back, its roar shaking his body. Raderstorf bounded up the steps just as the water enveloped the house. The deck to which he and his family now clung stood twelve feet above the ground, yet the water lapped up within a foot of the top. Entire buildings floated by them. After two minutes spent frantically searching for Carrie and her husband, Roger, they spied the couple paddling through the water—Roger one pinky finger poorer, but each of them safe. Amazingly, both families escaped serious injury.

Many died that day, yet Scott Raderstorf survived because his fear system worked to its absolute peak potential: within milliseconds of recognizing the threat, it threw his muscles into hyperdrive, commandeered his decision-making, and launched him into action. Had he paused even for a split second to think, had fear not instantly primed his body for a speedy escape, the tsunami would have overtaken him before he could reach safety. No

---

*In the 1960s, in fact, a group of scientists found that a massive dose of adrenaline can boost muscular power 30 percent above its normal limit.

wonder the Spartans paid such deep homage to fear; as Raderstorf could tell you, in the moments that truly matter, fear can become your closest ally.

For a very long time, we didn't know much about *how* this fight, flight, or freeze response worked in the brain—we just knew it did. The three *F*s are only the outward manifestations of a more complex system deep within, and although they tell us a lot about why fear exists, they reveal little about the strange emotional machinery that leaves some poised and others panicked in the face of danger. According to the New York University neuroscientist Joseph LeDoux, the physical sensations we associate with fear—pounding heart, trembling hands, quickened breath—are no more than "red herrings, detours, in the scientific study of emotions." To learn anything useful about fear, LeDoux continues, "what we need to elucidate is not so much the conscious state of fear or the accompanying responses, but the system that detects the danger in the first place." He should know. Through their innovative research, LeDoux and his talented colleagues in the "brain mafia" have revolutionized our understanding of fear by tracking down its secret lair in the brain: a tiny hub of cells called the amygdala. It's the hidden neural command center that explains all of fear's mysteries. It's also kind of a pain in the neck.

Prepare to meet your second brain.

Taking the Low Road

As a kid growing up in the Cajun country of Louisiana, Joseph LeDoux spent a lot of time contemplating meat. He had little choice: his father, Boo, was a butcher in the small bayou town of Eunice, and young Joe logged many hours at the family meat market, his developing brain exposed to a bounty of, well...*brains.*

Throughout LeDoux's childhood, one of his after-school duties was to remove slaughterhouse bullets from the brains of the newly arriving shipments of cattle. Far from making him gag, the job filled him with questions about how the brain—which, as LeDoux likes to point out, is just "intricately wired meat"—produces thoughts, perceptions, and especially emotions. This curiosity eventually propelled him up through the echelons of neuroscience, first at Louisiana State University and later at SUNY Stony Brook, where he and the eminent neuroscientist Michael Gazzaniga did pioneering research into split-brain cases—patients whose left and right hemispheres cannot communicate with each other. LeDoux hoped he could investigate the genesis of emotions in human subjects as well, but that path came with a frustrating hitch: since you couldn't exactly tinker with someone's brain (lest you should accidentally break it), you had to wait around for someone to suffer some exciting new form of brain damage and then rush off to study him. LeDoux was too impatient for that. He needed another way.

The answer was rats—many, many rats. In the early 1980s, soon after he accepted an assistant professorship in Cornell University's neurology department, LeDoux came across an intriguing new research method that Michael Fanselow was using to create a strong yet easily controlled emotional reaction in lab rats. It was called fear conditioning, and the protocol couldn't have been simpler. Step one: put a rat in a box. Step two: play a tone, like a loud beep. Step three: immediately after the tone, deliver an electric shock to the rat's feet through the floor of the cage. And *voilà*—in the same way that Pavlov conditioned his dogs to salivate at the sound of a dinner bell, the rats would now freeze in fear when you next played the tone, expecting another zap.* "This

---

*Incidentally, most of what we know about the science of fear comes from tormenting rats.

method turned out to be pretty convenient, because you had a discrete stimulus, the tone, that you could use to elicit a fixed emotional response, which was the freezing behavior," LeDoux told me. "These responses were very, very consistent and reliable, so it was a good way to study how the brain processed emotional stimuli." Now able to summon fear at will, with any luck LeDoux could hunt down its source within the brain.

There was just one problem: many of his colleagues thought such a task was impossible. When LeDoux sent out his first grant application to research fear in rats, he got rejected; the psychological establishment believed emotions were too nebulous to study. The thinking went something like this. By the 1970s and early '80s, technological innovations like CAT scanning began offering cognitive scientists firm proof that the human mind is an information-processing device much like a computer. The majority of this processing happens unconsciously, but we *do* consciously experience two kinds of mental activity: thoughts and emotions. Brain researchers had long known that cognition—the kind of rational, deliberative thinking that we use for analyzing problems and formulating plans—takes place in a large area in the front of the brain called the prefrontal cortex. Yet while cognition appeared cut-and-dried to brain researchers, emotions seemed baffling. Unlike rational thought, feelings were highly subjective and came in an infinite number of shades and textures. How could one possibly quantify feelings like hope, shame, and awe, or analyze the differences between, say, envy and greed? As LeDoux is quick to point out, we casually toss around dozens of terms—for example, *anxiety, terror,* and *apprehension*—to stand in for the umbrella concept of fear, but each word actually means something subtly different to us. It's hardly astonishing, then, that so many neuroscientists shied from studying emotions: they seemed like a never-ending labyrinth.

The vagueness of these emotional signifiers can be so confusing, in fact, that it's worth taking a moment to clearly define the three main terms we'll be using throughout this book: *fear, anxiety,* and *stress*. These three probably seem virtually identical, but to psychologists and neuroscientists they're quite distinct. *Fear* is the physical feeling you get when there's something dangerous in front of you *right now*, and its simple job is to get you to safety. *Anxiety*, on the other hand, is a cognitive phenomenon, and its purpose is to protect you from potential dangers that *might* pop up *in the future;* it's the free-floating sense of dread that often goes hand in hand with worry and unwarranted pessimism. *Stress* is perhaps the toughest term to pin down, but broadly speaking, it describes how our bodies respond to excessive demands: we feel stress when a situation grows emotionally difficult, when we have to work frantically to meet a deadline, when we endure too much pain, and so on. If the mental or physical demands of our lives overwhelm us, a stress reaction is the result. To put this all into context, imagine you're on a long train trip and you spot a man in your compartment who looks suspiciously like a wanted killer. If this man brandishes a gleaming knife at you, you will feel *fear;* if he does nothing but you find yourself fretting endlessly about what might happen, you will feel *anxiety*; and if he just turns out to be some jerk who wants to spend the next three hours telling you the entire plot of *Battlestar Galactica*, you will feel *stress*. Of course, fear, anxiety, and stress often overlap—fear can make you stressed, stress can make you more anxious—but we'll mostly talk about them as separate entities.

Joe LeDoux's target was pure, unfiltered *fear*, and he adamantly believed that the emotion's straightforward physiological profile meant he could employ the information-processing model of the brain to study it too—not just cognition. "My contribution at the beginning of this was to say that we can study the way the brain processes an emotional stimulus and produces an emotional response without making any assumption about how the animal experiences it consciously," LeDoux explained. With his tone-fearing conditioned rats, he could bring out the same fear reaction every time, and he never had to ask the rats to describe their subjective experience of the emotion. If some dedicated area of their rodent brains was processing the sound and triggering fear, he knew he could unearth it.

LeDoux pulled off this cranial Easter egg hunt through the process of elimination—which, unfortunately for his rats, literally entailed surgically eliminating different brain areas to see if the rats could still feel fear after they were gone. Just as information traveling through telephone lines will fail to reach its destination if a wire gets snipped, LeDoux assumed the sound of the tone would fail to elicit an emotional reaction if he disabled a crucial link in the fear circuit. Thus, he started with the tone's entry point, the rats' ears, and worked his way up the fear chain.

Almost immediately, he hit a paradox. When he destroyed a rat's auditory thalamus—the lower brain region that relays sound information from the ear to the higher-up auditory cortex—the rats no longer displayed any fear when he played the tone. This was no surprise: they were now functionally deaf. But when LeDoux lesioned the auditory cortex itself while leaving the thalamus intact, the rats still displayed fear *even though they couldn't consciously hear the tone.* This was somewhat mind-blowing. Without an intact auditory cortex you can't "hear" anything, since the cortex transforms external audio information into the conscious experience of sound, so how could the rats possibly freeze at a sound they didn't even know existed? If the rats were growing terrified at a noise they never consciously experienced, this meant that the auditory information from the ear must have split in the thalamus and traveled elsewhere in the brain—to some mysterious part of their subconscious mind that could still hear the tone. Amazingly, the fear system could bypass conscious thought entirely.

This new revelation presented a predicament: if the auditory information was branching off, it could be traveling *anywhere* in the brain. So to narrow down the suspects, LeDoux turned to one of neuroscience's James Bond–style tricks. Into each rat's auditory thalamus he injected a special tracer chemical that piggybacks on molecules that pass through the neurons in the region, leaving behind a map of its outbound connections. All LeDoux needed to do now was scare the rat, remove its brain, and then put it under a microscope capable of displaying dark field optics, which popped the chemical's path out into stark, luminous relief. In his 1996 book *The Emotional Brain*, LeDoux offers a description of this neural spectacle: "Bright orange particles formed streams and speckles against a dark blue-gray background. It was like looking into a strange world of inner space." The auditory thalamus, it turned out, sent neural projections into four brain regions, three of which had no effect on fear when LeDoux disabled them in his rodents. But when he dismantled the fourth region, his rats abruptly and completely lost the ability to become afraid. That region was the amygdala.

Now, it should be quite clear by this point that as one of the nation's top neuroscientists and an innovator in the study of emotion, Joe LeDoux is nothing if not a dignified emissary of science. Talk to him for just a moment, and your impression is of a man whose scorching gaze could reduce uncooperative graduate students to smoldering piles of ash. It comes as something of a shock, then, to learn that LeDoux's discovery about the amygdala's role in producing fear was so stirring to the hidden recesses of his soul that it has moved him to song—publicly, and often. With three of his NYU neuroscience colleagues, LeDoux formed a band with the goal of translating the miracles of brain research into rock and roll (or, as they've relabeled it, "heavy mental"). They call themselves the Amygdaloids, and the centerpiece of their oeuvre, an ode to the amygdala called "All in a Nut," may well be the only song ever penned about the brain's basal ganglia. "Why, why, why do we feel so afraid?" LeDoux croons. "Don't have to look very far / Don't get stuck in a rut / Don't go looking too hard / It's all in a nut…in your brain."

Well, technically speaking, it's all in *two* nuts: the left and right hemispheres both carry their own almond-shaped amygdalae, each of them situated about an inch and a half straight in from the temples. (The amygdala gets its name from the Greek word for "almond"—αμύγδαλο, or *amygdalo*—so describing it as "almond-shaped" is actually kind of redundant.) Taken together, these two neural clusters form the brain's hidden fear headquarters. The amygdala works just like a tiny security system lodged in our heads, and it's wired to help us detect and respond to danger as quickly as our mental hardware will allow—almost like an independent mind within the mind. Every second of every day, no matter if you're awake or asleep, your amygdala monitors all of the sensory information around you through its own special circuit from the senses, constantly searching for potential threats. Because the amygdala receives its raw data stream independently of the cortex, it scans things you don't consciously register, like background noises and objects in peripheral vision. It's kind of like having a batty Montana survivalist permanently camped out in your brain: the amygdala is vigilant, it's paranoid, and it has an itchy trigger finger. The instant that it detects a possible danger, it hits the physiological alarm switch, setting off a fight, flight, or freeze reaction within milliseconds.

The amygdala does its job so quickly, in fact, that it can fire up a fear reaction well before the conscious part of the brain registers that a potential threat is afoot—and thus, we really do react emotionally before we even have a clue what's going on. This phenomenon is best illustrated with another example. Suppose you're reclining in a comfortable chair in your living room, utterly absorbed in a gripping nonfiction book on, say, fear. You're so enthralled by the book's masterly prose that when a sudden draft blows a nearby door shut with a massive *SLAM!*, you leap out of the chair in terror. "Well, that was ironic," you think to yourself, your body flooding with relief, as you realize it was only the wind—nothing to worry about. For most of us, abrupt shocks like this are the closest we get to the fright Thag felt when he spotted the stampeding mastodon. But they're also puzzling. Why do we get spooked over something harmless? Shouldn't we be able to tell that it poses no threat?

Here's why things don't work that way. Recall that in the brains of LeDoux's rats, the auditory signal of the fearsome tone forked off in two directions, with one route blazing straight to the amygdala and another climbing up to the higher cognitive areas. LeDoux calls the first path the "low road": it's a lightning-fast superhighway to the fear center, but the information it transmits is raw and low in detail. The amygdala's job is to react to this brute data with an ultracautious, self-preserving response; consequently, when the door slams, the amygdala receives a basic report of a suspicious loud noise and makes you jump up, ready to fight or flee. The whole process takes just twelve milliseconds. Meanwhile, as the amygdala triggers a quick fear reaction based on the rough low road data, the same sensory information is wending its way through the "high road" to the cortex. This route gives us a much clearer picture of the situation, but the complex cortical networks run more slowly—they take thirty to forty milliseconds to do their processing, three times as long as the low road. Only now, after you've already responded to the threat, can the cortex send out the all clear, alerting the amygdala to the false alarm and returning it to a state of paranoid watchfulness.

Why, we might ask, is the high road so much slower than the low road? LeDoux explains the answer in terms of transit. "Let's say you're in the West Village of Manhattan and you want to get to the United Nations building," he told me. "You could go up to Forty-second Street and go across, and you'd pretty much be there. Or you could go up to Central Park and wander around, then go back down to Forty-second and across, and you'd also be there. Going the shorter route gets you there faster, but taking the longer route lets you gather more information along the way." The high road, then, is the mental equivalent of taking the scenic route, and because of the way the brain is designed, the low road will always be quicker. This is why Darwin couldn't stop himself from jumping back when the puff adder struck: the low road to his amygdala sparked a physiological reaction to the threat before the high road to his conscious brain even registered that the threat existed. Put another way, our fears are faster than our thoughts. We can't help this. No matter how hard you try to suppress it, you will always jump at the crash of the door and the snap of the snake.

Sometimes this high/low road system works to our advantage, and sometimes not. When you step off the sidewalk and the amygdala quickly spots an unfriendly bus speeding toward you from the side, the reflexive backwards leap it triggers will keep you alive. If a surprise party for your birthday sends you into cata-tonic shock, on the other hand, not so much. We might wish our higher cognitive machinery could keep up with the amygdala, but evolution, in its savage wisdom, knows that it's better to go through a thousand false alarms than to risk failing to react to a real danger just once. Because the low road has protected our ancestors so well for thousands of years, the mental technology of fear hasn't progressed much since Thag's time. As LeDoux points out, "In some ways we are emotional lizards."

LeDoux's breakthrough in tracing our feelings of fear back to their neurological source has revolutionized our understanding of the emotion, and his findings have been borne out in dozens of studies since, in human and animal subjects alike. When human patients have their amygdalae electrically stimulated during brain surgery, for instance, they become suffused with an intense feeling of anxious foreboding. And those unfortunate souls who suffer irreparable amygdala damage don't merely lose the ability to feel fear; they no longer even *recognize* the emotion. Ask them to describe a picture of a terrified face, and they can't tell you what feeling the person might be experiencing. Animals, too, display some truly peculiar behaviors without a fully functioning amygdala. The Emory University neuroscientist Michael Davis has conducted experiments on rats in which he disabled a rodent's amygdala and then placed it in a cage with its sworn nemesis: a house cat, dozing under the influence of a sedative. With their fear systems off-line, the usually wary rats transformed into rodent daredevils, scampering on top of the slumbering cat and nipping at its ears. Take away our ability to fear, and we become different beings entirely.

It's difficult to overstate the importance of understanding the amygdala, our skittish second brain, if we want to grasp fear's true nature. These neural clusters explain mysteries of fear that have vexed humankind for eons. For one, we now know why fear so often fails to respond to reason — that is, why you can spend hours telling someone who dreads flying about the airline industry's impressive safety measures, or about how it's far more dangerous to drive a car than travel by plane, yet their fear won't budge. The amygdala, brain researchers have found, shoots out a profusion of neural connections to the cortex, allowing it to supersede the conscious mind and immediately remove you from the path of that hurtling bus. But the neural conduits going in the other direction, from the cortex to the amygdala, are far fewer in number; in other words, our brains are actually designed to *thwart* our efforts to consciously override the fear response. (Evolution, you remember, has no faith in us.) It's not that the cortex grows completely helpless when fear arises, mind you—indeed, the amygdala and the cortex usually have a healthy reciprocal relationship, with each side's advantages able to balance out the other's flaws. The problem is simply that the amygdala can make it tougher for the thinking mind to work its magic.

Another riddle that the amygdala has solved is the question of why we become alarmed at mere *representations* of fearsome stimuli. If you plop a toy spider on the table in front of an arachnophobic, for instance, she'll scream out in horror, even if it's an obvious fake. The knowledge that the toy isn't dangerous won't dim her fear at all; she'll writhe in terror until it's gone. Why? Because the quick and blurry low road to her amygdala is providing a rough sketch of something alarmingly spiderlike—and since the amygdala's credo is "freak out first, ask questions later," this is all it needs to trigger fear.

For many of us, from those who deal with pesky little worries and phobias to those who struggle mightily with fear every day, the implication of this recent surge of amygdala research should come as nothing less than a revelation: what we *think* and what we *feel* really do live in different parts of the brain. We often chastise ourselves for our "irrational" fears, as though we're failing some kind of mental logic test whenever we become afraid against our will, but reason has nothing to do with it; *all* fear is technically irrational. No one is ever "crazy" for feeling afraid, no matter how strange the source of the fear, be it spiders or clowns or even antique Shaker rocking chairs.* Being anxious is never "your fault"; the worst thing we could say is that your subconscious fear system might be firing inappropriately, like a touchy car alarm that starts blaring when someone merely taps the fender. Learning to better handle the amygdala's ins and outs can be tricky, but as we'll see, it's an eminently workable problem.

Before we can delve into how to manage our assorted fears, however, we need to fill in one last piece of the amygdaloid puzzle. We now have a good grasp of how the fear response works, yet we also know that not all brains are created equal. Though we all fear some things universally—imminent tsunamis, ferocious bears, zombie hordes—we each harbor individualized fears as well. One person's amygdala calls in the cavalry at the sight of a toy spider while another person's fear system sees no danger whatsoever. Certainly, both people *know* on a conscious level that the plush arachnid is harmless, but their amygdalae have opposite reactions. So what causes these personalized fears? I've said already that the amygdala is a mind of its own, and I wasn't exaggerating: it even forms its own opinions about what you should fear. And frustratingly enough for us, those opinions are very tough to change.

Can't Get You out of My Head

Almost exactly one hundred years ago, a Swiss neurologist named Édouard Claparède performed a peculiar experiment that quickly became one of the best-known riddles in psychology. One of

---

*You probably thought I was making that last one up for laughs, but the actor Billy Bob Thornton has actually admitted to having a massive phobia of antique furniture. Not that I'm holding Mr. Thornton up as a paragon of mental health.

Claparède's patients, a forty-seven-year-old woman, suffered from a rare and extraordinary form of brain damage that left her reasoning powers and pre-injury memories intact but completely disabled her ability to form *new* memories. It was like a reverse case of amnesia: she recalled everything about the past but nothing about the present. The patient could carry on a conversation and hold on to short-term memory for perhaps ten minutes, yet beyond that point all information disappeared into some neural void, never to return. Every day, Claparède met with this woman to check on her condition, and every day, he had to introduce himself to her anew. If he left the room for twenty minutes and came back, she looked at him as though they'd never met.

For one of these meetings, Claparède decided to vary the routine with a little test. Before entering the room, he hid a pin in his palm, its edge pricking out. As soon as the woman shook his hand, she recoiled in pain and shock—but hey, she wasn't going to remember it, so no harm done, right? When Claparède came in the following day and stuck out his pinless hand for her to shake, however, the woman now refused to touch it. She couldn't explain her reluctance, nor did she have any memory of the day before, yet she claimed the prospect of shaking his hand filled her with dread. No one could fathom it: if she was unable to form new memories, how could she know Claparède's handshake might be hazardous? The psychologists of the era posited that some hidden "implicit" memory system must be at play here, but it would be many years before neuroscience could offer a firm answer to this fear-memory puzzle.

Our fear system relies heavily on learning and memory to help it do its job, after all. In order to prevent us from seeing everything around us as a life-threatening hazard—"Oh no, a kitten! Good God, there's a banana!"—the amygdala needs a quick and reliable way to tell the difference between harmless things and dangerous things. So in the same way that your computer's antivirus program compares each file on your hard drive with its data bank of malicious software, your amygdala scans all incoming stimuli against a memory bank of threats. If it gets a close enough match (say, a scurrying black critter), it fires up a fear reaction ("Spider alert!"). Memory, then, is an essential ingredient in fear. When we say that Claparède's patient developed a fear of shaking his hand after her first prickly encounter, what we mean is that her brain learned a new entry for its threat database: perhaps "outstretched palms of smart-ass doctors." The same is true of LeDoux's fear-conditioned lab rats. Before his experiment, the rats were indifferent to the digital tone, but after just a single tone-shock pairing they instantly learned that tones meant unpleasant jolts and filed the information away for future use. Because Claparède's memory-impaired patient was still able to form a new fear, the information about the pin-prick clearly couldn't have gone to her brain's all-purpose memory center. This neural structure, the sea-horse-shaped hippocampus, is the region that's responsible for helping you remember important facts like the date of your wedding anniversary, your online banking password, and where you parked your car.

Oh, wait a minute—you *can't* remember those things, can you? In fact, you forget them constantly; you're probably wandering aimlessly around a parking garage *as you read this*. This is why fear memories don't reside in the hippocampus, the brain's home for most long-term memories. Our factual, or "explicit," memories have a troubling tendency to weaken over time (really: try to recall what you had for lunch last Tuesday), but unless we want to get eaten, we can't afford to have the memory of a threat to our safety fade. If you're a rat, for example, you never, ever want to forget that a house cat is a dire threat; it should be engraved forever in your little brain. To keep us safest, fear memories have to be deep-seated and inflexible, and they have to be "implicit"—that is, they have to work beyond our awareness. (Implicit memory lets us perform memorized actions without conscious effort, like touch-typing or driving a car.) So there could be no more logical or convenient place to store these unique, etched-in-stone emotional memories than in the seat of fright itself: the amygdala. Within those two little clusters lie the keys to our deepest, darkest fear memories, from humankind's most ancient terrors to the new connections we make all the time.

Let's start with the primal fears. When we're born, the amygdala doesn't start off as a blank slate, merely awaiting instruction on which things to fear. Mother Nature was kind enough to preprogram our brains with innate aversions to things humans generally wanted to avoid over the ages: snakes, spiders, heights, darkness, sudden loud noises that might indicate an attacker, closed-in spaces.* Each of us expresses these fears in different proportions, but they never have to be learned from experience. At birth, healthy infants immediately show what's called the Moro reflex: if you let a baby's head suddenly fall backwards, the infant will display its inborn fear of falling, flashing a startled look and splaying its arms out sideways as if to catch itself. Similarly, even on islands that have no snake species slithering about—and thus where the indigenous inhabitants never had to worry about snakebites—native kids will still shrink in terror when they first see a real snake. Somehow, snake fears are encoded in human DNA.

---

*Many who have a fear of small spaces (claustrophobia) also fear open spaces (agoraphobia), and some speculate that the two come from the same source: a primitive fear of being unable to escape a sudden threat. Caught in an open field or cornered in a tight space, we have no way to flee or hide from a predator.

(Just ask LeDoux, who harbors a substantial snake phobia, despite living in the relatively serpent-free confines of New York City.)

As the science journalist Rush W. Dozier points out in his useful book *Fear Itself,* repression-based Freudian interpretations of these preloaded fears dominated psychology for many years, often to bizarre effect. Wrote one psychoanalyst of arachnophobia, "The spider is a representative of the dangerous (orally devouring and anally castrating) mother, and...the main problem of these patients seems to center around their sexual identification and bisexuality." This proved somewhat confusing to the phobics themselves, who thought—just like their ancestors—that they were afraid of spiders because spiders can kill you, not because they needed to ask Mom to stop with all the anal castration. But the truth is, our innate fear of these primal threats is so strong that we're wired to identify them with astonishing quickness. In a 2008 study, two University of Virginia psychologists showed subjects a grid of nine images on a computer screen and asked them to do one of two things: find a snake among eight nonthreatening items, like flowers or frogs, or find a nonthreatening object among eight snakes. The subjects each picked out the snake image amid the harmless items far faster than they could identify the harmless item among the snakes; it seems we're primed to spot these potential hazards before any other stimuli.

The world contains a greater variety of threats to our well-being than this primitive menagerie of creepers and crawlers, however, so evolution also endowed us with a system for rapidly assimilating *new* threat information. Fear learning, explains Michael Davis, the Emory psychologist, is arguably the strongest memory system we have. "The interesting thing about fear memories is that you can learn them instantly and they last a lifetime," he told me. "There isn't anything else that you learn instantly. Even something like your mother's name, you've had many rehearsals of that, hearing it and writing it down. But what's unique about fear conditioning is that one thing happens to you that is potentially life-threatening and you'll never forget it. That's very adaptive, but it's also a problem." Here's why the instant learning is adaptive. Consider our caveman friend Thag once more: if he encounters an enraged mastodon, he's going to be lucky to survive the experience just once. Ideally, then, his brain will stockpile as much information about the encounter as possible, enabling him to recognize the warning signs of another attack before it happens. So, always looking to maximize our safety, the amygdala quickly etches the rough details about the threat into its data banks so he'll immediately go on alert when he next stumbles into a similar situation.

And this is precisely why fear learning can be a problem: the system is imprecise, but the fear memories the amygdala lays down are deep and persistent. With a single bad experience, we can become conditioned to fear things that are totally harmless. A man who gets in a horrific car accident won't just grow skittish about driving; he might also become averse to yellow cars like the one that hit him, or to leather seats like the one in his car. Clearly, none of these details had anything to do with his accident—it's simply another case of the rough-and-ready amygdala erring on the side of caution. Fear is a sledgehammer, not a scalpel: it goes for power over precision.

Luckily, the amygdala finds it tougher to form fears about innocuous things than about legitimate threats. In the 1980s, the psychologist Susan Mineka performed a study with baby lab monkeys to test their predisposition to form fears. Monkeys don't share our strong innate fear of snakes, so when Mineka showed the young primates a live snake, they weren't especially perturbed. When she showed the monkeys a video of another monkey howling in terror at a snake, though, the monkeys quickly began to ape (sorry) the fear reaction; this process is called modeling, and it's a way for us to learn a fear without experiencing it ourselves. (If your monkey friend is freaking out about it, you don't need to find out firsthand that it's unpleasant, right?) But when Mineka superimposed an image of a flower over the snake in the video—so that it appeared the monkey was panicking over a rose—the baby monkeys developed no aversion to the flower at all. Forming a snake fear is effortless, but it takes a lot more to make them afraid of something as harmless as a flower. It's the same with us. Writes Dozier, "A child can easily become afraid of snakes, but no matter how ingenious and persistent the researcher, it is almost impossible to make him afraid of a toy duck."

Note that Dozier said *almost* impossible, because as the famed Johns Hopkins University behaviorist John B. Watson proved in a troubling 1920 experiment, we really can learn to fear anything—if the stimulus is made sufficiently terrifying. Since this trial predated the concept of ethics, Watson took as his subject an eleven-month-old infant, who has since been known as Little Albert. Little Albert was a placid, happy child who was cared for during the day at Baltimore's Harriet Lane Hospital, where his mother was a nurse. (Appallingly, Albert's mother evidently had no clue her son was being used in Watson's experiment.) For his fear stimulus, Watson opted to use cute animals. When he first presented Little Albert with a snowy-coated rabbit or lab rat, the infant would gurgle with pleasure, as he was a natural enthusiast of fluffy, soft things. But then Watson began his conditioning. Now, every time Little Albert touched a white rat, Watson bashed a hammer into a steel bar behind the infant's head, the earsplitting sound sending him into a state of terror. After a few exposures, the conditioning was complete: Little Albert squealed in horror at the white rat's arrival, hurriedly attempting to crawl away. Consistent with the amygdala's blurry vision problem, Albert's fear generalized to all furry things; he now grew distressed when he saw dogs, fur coats, even a Santa Claus beard. Watson had planned to decondition Little Albert's fear after the trials concluded, but as soon as the child's mother found out about the experiment, she whisked her son away, telling no one where she was headed. Somewhat incredibly, Watson later went on to pen popular guides to child rearing. To this day, no one knows what happened to Little Albert.

In all likelihood, though, his terror lasted quite a while. According to Fanselow, the UCLA fear expert, fear conditioning can persist over an animal's lifetime, a fact he's proven in his own lab. "It's actually the world record for holding a memory without any memory impairment," he explained. "We took adult rats and did fear conditioning with them, and then we waited a year and a half. That might not sound like a long time, but it's basically the entire adult life span of the rat. And when we tested them again with the tone, there was absolutely no forgetting of the fear. Not one bit." This is how efficient and robust the brain's fear system is. You can overcome a fear—a process we will explore soon enough— but like the proverbial elephant, your amygdala never forgets.

Breaking the Loop

Both in his escape from the tsunami and in its psychological aftermath, Scott Raderstorf was lucky. When he tells the story of his terrifying footrace against the churning wall of water, he does so with a mixture of awe and humor. "I've read guides to natural disasters since it happened, and the one that was most telling said that for tsunamis, there's really nothing you can do," he said with a laugh. "If you're close enough to see a tsunami, it's too late." Raderstorf knows he's fortunate to be alive, but the experience failed to traumatize him—in fact, the tsunami didn't even spoil the family's vacation. Because the Thai navy evacuated women and children first, Scott got separated from Joellen and the kids for a few days, but soon enough they were trekking across the globe once more. The only lingering psychological residue he carries is a recurring dream of trees and houses breaking apart under the force of the water—specifically the unnatural, low cracking noise they made, like the sound of bones breaking. Most others wouldn't emerge so unscathed.

But Raderstorf still remembers the event as clearly as if it happened yesterday, which is common among those who go through such harrowing experiences. In part through the influence of the amygdala, emotions often amplify our memories so that the most affecting moments of our lives—both good and bad—stand out in clear relief, like mental photographs. Psychologists call this the flashbulb effect. Everyone, it is said, remembers exactly where they were when they heard that President Kennedy had been shot, and the same is true of all tragedies, from Pearl Harbor to 9/11. (But although we're frequently *confident* that we recall a flashbulb moment perfectly, we're not always *right*. In one study, psychologists asked students to recount their memories of the explosion of the space shuttle *Challenger* within a few days of the event, and then followed up with them a few years later; on the second date, the students recalled the explosion in vivid detail, but their memories were now fraught with inaccuracies.) Sometimes, these intense, emotionally charged memories become far more psychologically troublesome, as we see in the deep trauma of many combat veterans. In cases of post-traumatic stress disorder, or PTSD, fear and memory enter into an awful feedback loop. The vibrant memory springs out at slight provocations like the sound of a car backfiring, and the amygdala reacts to this detailed recollection as if it's actually happening again, right now.

This is the fear mechanism at its most troubling: when our

ever-vigilant security system seems to turn against us, sending us into flurries of fright over invented dangers. Even if we never go through the kind of life-threatening experiences that soldiers routinely face, each of us builds up our own unique catalog of aversions over the years. Whether it's a dread of public speaking, a terror of flying, or a nagging tendency to worry about relationships or job security, the amygdala often develops fears we don't want and fires away just when we'd like to feel relaxed and composed. Because we tend to see fear and anxiety as unwelcome hindrances, we often assume that we can't exhibit grace under fire until we've driven them away. Fear and cool strike us as oppositional forces, like darkness and light; if you're afraid, then you aren't showing courage or poise, end of story.

But nothing could be further from the truth. Fear and cool are far more compatible, and even *essential* to each other, than you might think. Our anxieties never need to be our enemies; indeed, some of the most neurotic and fearful people on the planet are also the iciest customers under fire. What truly separates the cool-headed from the hotheads in tense times isn't *whether* they feel fear—which is largely beyond their conscious control—but how they *relate* to their fear. Over the following chapters, we will survey two main landscapes: the hidden dynamics of fear, anxiety, and stress; and the secret ingredients that keep people poised through daunting situations. While these might seem like separate streams, both of them flow into the same river. Developing true cool isn't just a matter of reducing the amount of fear we feel, nor is it a matter of applying a set of rules to attain perfect poise. It's a matter of learning the right way to be afraid.

Now, at some point in the future, neuroscience may well provide a way to avoid fears and traumatic memories altogether. In 2002, for example, the Harvard University psychiatrist Roger Pitman conducted an experimental trial on recent accident victims, using the anti-hypertension drug propranolol to curb their formation of fear memories. Pitman and his team approached forty-one patients admitted to the emergency room for mishaps like car accidents, and within six hours of the event, he randomly assigned them to take either a placebo or the propranolol every day for ten days. Propranolol blocks the action of adrenaline—the same hormone that fuels the body's fear system—and Pitman reasoned that if adrenaline couldn't do its job, the accident victims' memories wouldn't receive the same traumatic charge. Sure enough, three months later, the patients who took propranolol had significantly fewer trauma symptoms than those who took the placebo. Taking this idea to a more surreal level, Joe LeDoux and his colleagues are now sketching out a method to selectively *annihilate* emotional memories. When you inject the brain with a chemical that inhibits protein synthesis, he explains, any memory you retrieve will fail to reconsolidate—the chemical makes the memory unfurl, as if you untied a neural knot. In the lab, LeDoux has treated his rats' brains with such a substance and eradicated their fear of the alarming digital tone after just one exposure; days later, the rats showed no reaction to the tone at all. Used carefully, this procedure could be a boon for trauma victims, though some of the inquiries LeDoux has received about it have left him nonplussed. One man who came across his research called to ask, with the most fervent hope, if LeDoux could erase the memory of his ex-wife.

Until these cutting-edge therapies hit your local psychologist's office, however—and don't hold your breath quite yet—we mostly have to deal with fear the old-fashioned way. And therein lies the rub: it's a tough task. Bad experiences etch fear memories into our brains against our will, and our amygdalae trigger fright when we want most to be unruffled. It's not the most user-friendly system in the world, because it wasn't intended to be fun to operate—it's there to keep us alive. Yet most of the time when we feel anxious or stressed, it's not about some mortal danger; we're reacting to the tension of a math test or the pressure of a last-second basketball shot. Instead of helping us thrive, fear seems to get in our way, and we find it difficult not to see it as thwarting our chances for success.* We'd like to be able to turn off the fear reaction by flicking some sort of psychological light switch, but that's not the way the system is rigged. Fear is an incredible, powerful, truly lifesaving mechanism, yet we all struggle with its idiosyncrasies.

Luckily, as impossible as this system is to control, our fears are not necessarily our fate. No one is predestined to obey the whims of the amygdala. Adapting our behavior and beliefs can bring about huge changes in how the amygdala expresses fear; we can even help our brains learn to inhibit the fear response before it ever gets off the ground. This, too, is a neurological process that researchers are only now beginning to understand, but one finding is abundantly clear: we can get over virtually any fear if we approach it the right way. Of course, stepping onto this path can be a bit tricky—but not because approaching fear wisely is complex or unnatural. No, the biggest hitch in getting over our fears is that these days, we've fallen into the baffling habit of addressing them as *unwisely* as humanly possible.

---

*We'll get into how fear *can* help us succeed at these sorts of things in Chapter Four.

CHAPTER TWO

The Worry Trap: Eight Awful Ways (Plus a Few *Good* Ways) to Deal with Fear and Anxiety

On the great cosmic list of things we'd all like to avoid, illness ranks right near the top. Nobody wants to get sick. So, worrying about our health from time to time is perfectly natural, and even occasionally beneficial—for instance, if health anxiety keeps you from renting a vacation cottage in a scenic bubonic plague quarantine zone. Yet there are some among us who take their medical fears to dizzying heights, becoming so captivated by health worries that they seem to inhabit a skewed alternate universe where every headache and hangnail signals impending doom. Their fretful realm is a peculiar place to visit, but it's an enlightening trip for us to take. Because when it comes to case studies in how *not* to deal with fear and anxiety, there is no better place to begin than in the worried world of the devoted hypochondriac.

We begin our tour of this parallel universe at its virtual town square: the Internet's network of hypochondria message boards, where every day thousands of health-anxious citizens catalog their symptoms and trade worries with fellow residents. Within these digital confines, no complaint is too vague or trivial to warrant a full-blown freakout. In my perusing, I've come across posts titled "armpit pain—please respond!!"; "receding gums = certain death?"; and "fast growing fingernails...help!" Anxious writers track their heart rate and body temperature minute by minute, as though they're monitoring a nuclear reactor on the verge of meltdown. Ominous symptoms coincidentally appear after reading a news story about a rare disease, or after watching a hospital drama like *House*. (Multiple sclerosis is a popular self-diagnosis, since the symptoms, shakiness and muscle tension, for example, mimic those of anxiety.) Here, the standard rules of reason fade away. One college-aged girl wrote in terror about a black dot on her gums, convinced the end times had arrived. "Slight overreaction: it was just a piece of basil," she reported moments later—yet she still soldiered on, undaunted, with new details about a probable hip tumor. A visitor to this world might be tempted to think these people are joking, yet the writers are legitimately convinced that they're gravely ill.

Arthur Barsky, a Harvard Medical School psychiatrist and hypochondria researcher, explains that these health worriers have tremendous difficulty believing that their random physical sensations might be harmless. "They're not faking their symptoms," Barsky told me. "It's just that there's no serious medical illness that explains them. There are all kinds of benign reasons for having symptoms—not getting enough sleep, dietary reasons, stress—but they can't accept that it's something trivial." In other words, their illness is one of uncertainty: faced with a vague physical sensation, the hypochondriac dwells not in the *probability* that it's benign, but in the *possibility*, however slim, that it heralds something catastrophic. Trifles balloon into looming catastrophes. Worries explode into certainties. Hypochondria, then, is anxiety run amok. Unable to tolerate the dread of uncertainty, the hypochondriac gives fear free rein over his life. "The more sensitive call hypochondriacs the 'worried well,'" writes Jennifer Traig in her hypochondria memoir *Well Enough Alone*. "The name is apt. We do, indeed, worry extremely well. We worry consummately and constantly." Over and over, hypochondriacs' health anxieties spark behaviors that don't just fail to ease their fears—they actually make them worse.

As an example, consider one weapon in the hypochondriac arsenal: seeking reassurance. Plagued by dire health worries, many hypochondriacs try to allay their anxiety by embarking on epic medical quests to determine, with total certainty, that they're okay. (An estimated 5 percent of all visits to primary care doctors in the United States stem from hypochondria, at an annual cost of $20 billion to the health care system.) But this tactic has a major flaw. Hypochondriacs never feel reassured, because no physician or scanner on earth can deliver a diagnosis with perfect precision—there's always a chance that it's wrong. "They're searching for absolute certainty that they are not ill, and that's impossible," said Barsky. "You can have a total body scan that's normal now, and then have a heart attack ten minutes later." Still, they keep searching for reassurance that never comes, scheduling new tests, requesting second and third opinions, feeding the anxiety. This pattern has earned hypochondriacs the eternal enmity of doctors, who call them "gomers," short for "Get out of my emergency room." Physicians grow frustrated not only about hypochondriacs' constant backseat doctoring and "organ recitals"—the litany of nebulous complaints about every square inch of their bodies—but also about the fact that nothing they say helps. Explained Barsky, "The doctor thinks, 'Everything I tell you just makes you more anxious.'"

Because of behavior like this, hypochondriacs get little respect. Most often, they get laughed at. Take Henrietta Darwin, beloved fourth child of Charles Darwin. In the Darwin household, hypochondria was an art, and Etty was the Picasso of the form. When she was thirteen, a doctor who examined her for a mild fever advised her to have breakfast in bed for a few days; as her niece later recounted, *"She never got up to breakfast again in all her life."* Lounging about the house, Etty draped a silk handkerchief over her left foot, since it always seemed a shade colder than her right. But Etty's true neurotic masterpiece was an improvised gas mask, which she donned when company came calling. "It was an ordinary wire kitchen-strainer, stuffed with antiseptic cotton-wool, and tied on like a snout, with elastic over her ears," recalled her niece. "In this she would receive visitors and discuss politics in a hollow voice out of her eucalyptus-scented seclusion, oblivious of the fact that they might be struggling with fits of laughter."* Some hypochondriacs do the teasing themselves, as the health-anxious humorist Gene Weingarten demonstrates in his description of another unhelpful behavior: symptom testing. "The ordinary person will notice a slight spastic tugging at his eyelid . . . and go 'Hmm,'" he writes in *The Hypochondriac's Guide to Life. And Death.* "A hypochondriac would not go 'Hmm' unless you told him there was a new fatal disease whose first symptom is the inability to say 'Hmm.' Then he would say 'Hmm' 1,723 times a day until he got laryngitis and could no longer say 'Hmm,' which would of course constitute proof that he is dying."

The true scourge of the hypochondriac, however, is a newer addition to his tool kit: the Internet. In darker ages, hypochondriacs who wanted to research arcane diseases had to trudge to a library and pore over musty medical reference texts. Now, with

---

*The most legendary hypochondriac in history, though, was the poet Sara Teasdale, who even employed her own full-time nurse. One day in 1933, a blood vessel burst in Teasdale's hand, which the poet interpreted as a sure sign of an impending stroke; convinced of her grim fate, she decided to skip the formalities and take a massive overdose of sleeping pills. The coroner later reported that she'd been in perfect health.

websites like WebMD and MedicineNet, these resources are never more than a click away. This has been an exquisitely unhelpful development for the worried well; if knowledge is power, the Internet's information overload makes the anxious mind short-circuit. One needn't even be a full-on hypochondriac to fall victim to this "cyberchondria." In 2008, for instance, researchers at Microsoft performed a startling study of their employees' health search habits. Their results showed not only that more than half of their employees had significantly interrupted their day at some point to hunt for medical information, but also that these searches led them to the gravest possible diagnosis. Test it yourself. Google "leg pain" and on the first page of results alone, you'll find information about deep vein thrombosis, diabetes, restless leg syndrome, herniated discs, blocked arteries, and fatal heart conditions, plus a link to "193 disease causes of leg pain." Is it really a mystery why the Internet has sent hypochondria rates climbing?

At first blush, hypochondria might seem an odd topic to delve into if we want to understand how to manage fear, stress, and pressure. After all, could anyone possibly be *less* cool-headed than the person who assumes every itch foretells cancer? True hypochondriacs appear almost deranged in their slavish obedience to their anxieties, which drive them toward actions that are mind-bendingly counterproductive. Seeking reassurance that never arrives, worrying endlessly about medical catastrophe, researching exotic diseases, testing symptoms until they worsen—these tendencies are so self-destructive that we ask, *Why do they even keep doing them?* And the answer to this question is the crux of it all, not just for hypochondriacs, but for every one of us: we think our worries help guide us, yet more often than not, anxiety tells us to do precisely the wrong things. Hypochondriacs provide us with a valuable object lesson, because, in reality, they're more like the rest of us than we'd like to think.

The Eight Mistakes

Though we typically envision them as being one and the same, our friends fear and anxiety have vastly different personalities and neural profiles. Think of them as fraternal twins; while fear and anxiety resemble each other and share the same parent—the amygdala—they're also distinct individuals, with their own tastes and habits. As we discussed in the last chapter, fear is mainly *physiological*, yet anxiety is predominantly *cognitive:* fear supercharges the body to escape real danger right now, and anxiety motivates the brain to figure out how to avoid theoretical danger in the future. One screams, "Run away!" while the other ponders, "What if?" Fear can cause anxiety and anxiety can spark fear, but they also work independently of each other.

In 2007, an amusing experiment at University College London clearly outlined the neurological differences between the two states and showed how we can slide freely between them. For this study, researchers monitored brain activity scans as subjects played a Pac-Man-style video game with a painful twist: if the red villain chasing them through the virtual maze caught up with their character, the human player received an electric shock. (Psychologists are never truly happy unless they're administering electric shocks.) When the players had the monster at a comfortable distance, the scanners showed high activity in the prefrontal cortex; the subjects were planning their next moves to avoid the threat, consistent with a state of anxiety. But as the predator drew closer and closer, the action in the brain shifted. Blood flow drained from the prefrontal cortex and surged in the more primitive lower brain areas involved in the fight, flight, or freeze reaction, making the players more impulsive. The danger was now immediate, and fear took over the show. This experiment shows us how our onboard fear and anxiety systems form a continuum; we get *anxious* about a distant threat as we plot how to evade it, and then we glide toward *fear* when the threat looms nearer. Because fear and anxiety are part of this single sliding scale, the same advice, good and bad, applies to both characters.

Fear is something that all animals experience, but as far as we know only humans, with our expanded cortical regions for thinking and planning, can feel anxiety. Unless you've got some particularly bizarre habits—say, dodging highway traffic for fun—chances are that you feel more anxiety than fear in daily life. Indeed, with our soaring rates of anxiety disorders (one in three will develop one at some point), Americans are arguably the greatest worriers the planet has ever known. But fortunately, while anxiety is uniquely human, we're also the only animals who can deliberately seek to overcome our fears. (When you turn on the vacuum cleaner, your tense house cat doesn't think, *I appear to be afraid—perhaps therapy is in order?* He runs away.) This dampening effect happens through a neurological process called extinction, and it's the route through which any of us can quell our fears and ease our anxieties.

There's just one problem with revving up this extinction system and processing our fears: we have a tendency to sabotage it before it even gets off the ground. As we've seen already with hypochondriacs, when we follow anxiety's instructions about how to become safe, we typically end up feeling worse, not better. Anxiety, for all of its good intentions, often leads us toward a set of errors, dead ends, and red herrings—eight of them, to be precise.

Mistake #1: Being born that way

Okay, since you can't really tinker with your genetic heritage, this technically isn't a *mistake*, but it does provide the foundation for how we all experience fear and anxiety. When the subject of cool-headedness pops up in conversation, the first question that anyone asks is usually this: Are poised people *born* or *made*? Do they emerge from the womb with sanguine looks on their faces, ready to perform lifesaving surgery in the next hospital room if necessary, or does their coolness come from life experience?

In 2001, after conducting a comprehensive study on the genetics of phobias, the Virginia Commonwealth University psychiatrist Kenneth Kendler supplied us with our best answer yet. Kendler and his colleagues surveyed 1,200 pairs of male twins, some fraternal and some identical, looking at each subject's individual phobias. Because the twins all had the same upbringing and early environment, yet only the *identical* twins had the same DNA, Kendler could contrast the phobia data from the identical pairs with the fraternal pairs and filter out the environmental influence altogether, arriving at a pure figure for genetic susceptibility to anxiety. "The bottom line is that the genetic influences on these phobias are clearly present and moderate in strength," Kendler told me. "That translates into about thirty percent of the individual differences in our vulnerability to anxiety." In other words, around a third of your tendency toward anxiety lurks in your DNA. But Kendler's research also yielded another, more surprising finding: our individual fears often come bundled in our genes, too. "You'd think that most people would be scared of rats because they'd been bitten by a rat, but this turns out not to be empirically true," he explained. "In fact, most people are scared of rats without ever having been close to one. The fear is more likely to be innate."

To this genetic stew we can add another risk factor that affects fully half of the human race: being female. According to Kendler, women are nearly *twice* as prone to anxiety as men—and as you might imagine, the debates over why this is the case can turn combustible. Explains Michelle Craske, a renowned psychologist and anxiety researcher at UCLA, the question appears to hinge at least in part on the differences in how we raise our kids to display emotion. "From a socialization angle, there's quite a lot of evidence that little girls who exhibit shyness or anxiety are reinforced for it, whereas little boys who exhibit that behavior might even be punished for it," Craske told me. Call it the "skinned knee effect": parents coddle girls who cry after a painful scrape but tell boys to suck it up, and this formative link between displays of negative emotion and kisses from Mom may well predispose girls to show greater anxiety later in life. No one knows the precise answer yet.

Truth be told, we know that anxiety is 30 percent genetic, but we're still murky on the other 70 percent of the equation. All we can say for sure is that many of us are born with worry wired into our brains. Yet our genes are not necessarily our destiny. "The metaphor I use is that life is like a symphony," Kendler said. "You can change the themes you get from your genes a bit, though not a lot. But how you play out those themes over your life, that's more under your control."

Mistake #2: Seeking unattainable certainty and control

It's time for a thought experiment. Imagine that you've been beamed *Star Trek*–style into a woodland glade—a situation that seems pleasant enough until you notice a bear nosing around nearby, oblivious to your presence for now. In this iffy predicament, you'll feel not just pure fear (in the form of a flight-enabling burst of adrenaline) but plenty of anxiety as well: *What if it notices me? What if it attacks? What if it has access to the* Star Trek *transporter?* Like Darwin at the snake cage, you can't really help your innate fear reaction, but there are two ways that your worry-laden anxiety could end immediately. One way would be to gain *control* over the situation. If a laser rifle labeled "Bear Vaporizer" suddenly materialized in your hands, your new sense of control would basically end the what-if process completely; you'd always have the power to eradicate the bear threat. The other way would be to attain a feeling of *certainty* about the future. Suppose that God himself poked his head out from the heavens and said, "I did a pretty great job with bears, eh? Scary stuff. Anyway, don't worry—this bear definitely won't attack you." With the certainty that you'll be safe, your anxiety would likewise disappear. Thus, your original bear-related anxiety came from two sources: lack of control, and uncertainty.

Especially uncertainty. Put simply, uncertainty is anxiety's archenemy. It is the primary spring from which worry and apprehension flow, so much so that we can think of anxiety itself as a drive to eliminate uncertainty. For example, the anxiety researcher Michel Dugas has found that for chronic worriers, the single greatest cause of their anxiety is their refusal to accept the possibility, no matter how small, of a negative event happening in the future. (We've seen this at work already with hypochondriacs.) A strong intolerance for uncertainty actually changes the way we perceive the world. People who can't stand uncertainty interpret ambiguous information not as vague or neutral, but as *threatening*; they loathe uncertainty so much that, according to the cognitive therapist Robert Leahy, research has shown they "would rather know a negative outcome for sure than face the possibility of an uncertain outcome that could be positive." Our anxious drive to gain a sense of control comes from the same source. Having control over a situation gives us a feeling of certainty about the future, and we no longer have anything to worry about. But lacking control just makes us more uncertain—and consequently more anxious.

Certainty and control, then, are something of a mixed blessing. We're less anxious when we feel we have them, yet seeking them when they're not feasible—trying to control the uncontrollable or find total certainty in an uncertain world—will only make things worse. Hypochondriacs who quest after an irrefutable guarantee of health, for instance, just end up locking themselves in an endless cycle of worry. An essential element in managing anxiety, then, is learning to recognize when life is inherently uncertain or out of our control and getting used to that fact instead of fighting against it.

Mistake #3: Worrying about it

In the early 1970s, the psychologist Tom Borkovec kicked off his research career with an investigation into insomnia, which the experts of the day believed to be a product of too much physiological arousal. Borkovec's research crew assumed that teaching their insomniac subjects physical relaxation techniques would yield an easier transition into sleep, yet this tactic didn't pay the dividends they expected; physically tense subjects might slip off easily, while calmer people might lie there staring at the ceiling. The more important variable, they soon found, wasn't *physiological* activity but *cognitive* activity. The insomniacs were fretting, ruminating, and generally worrying themselves awake. This insight sent Borkovec down a long and productive path into the deepest workings of worry, and what he and his collaborators have found might surprise you: worry isn't just unpleasant, repetitive, and useless—it also *hinders* us from getting over fears.

Evelyn Behar, one of Borkovec's collaborators who now teaches at the University of Illinois at Chicago, defines worry as a "ruminative linguistic activity about potentially threatening future events," and the operative word there is *linguistic*; worry consists almost entirely of talking to ourselves in our heads. Our common assumptions about this nagging internal chatter are almost comically out of whack. "Some people have positive associations with worry," Behar told me. "They believe that worry helps them prepare. But worry doesn't serve people quite as well as they think." For example, one of the strongest findings in worry research is that fretting almost never leads to actual solutions to problems. We feel like we're figuring out an issue, but we're really just worrying in circles. In part, this is because we tend to fret about things that are, for lack of a better word, *bullshit*. "Another thing we know from having people fill out worry diaries is that ninety-five percent of worries never come true," explained Behar.* "And with the things that *do* come true, they end up coping much better than they ever expected." One would think that if (a) worrying rarely yields useful results, (b) worries almost never come true, and (c) we handle the things we worry about just fine, then we'd all stop worrying forever. But we don't, for two reasons. The first reason is that the lack of negative outcomes *reinforces* the worry habit. Because feared events almost never follow worry, we start subconsciously believing that worry prevents such things from happening.

The second reason, though, accounts for why worry is so toxic to the fear extinction process: worry actually *mutes* emotional expression, which makes it tougher for us to overcome a fear. When researchers show subjects a scary image (say, a gory car accident site), their physical level of fear spikes, which is precisely what you'd expect. When they start worrying about that topic, however, even though they report feeling highly anxious, their overall bodily arousal level drops; the verbal worry loop buries their immediate fear. People get over fears only when they expose

---

*If your reaction to that sentence was "Oh my God, that means five percent of my worries are going to come true," then congratulations: you're a serious worrier. Welcome to the club. Membership materials are on their way.

themselves to those fears (which we'll explore in depth in a moment), but worrying seems to act as a mental buffer against facing what troubles you, "incubating" your fear instead of processing it. As Borkovec and his coauthor Lizabeth Roemer have written, worrying has the short-term payoff of making us less afraid, but it traps us in a cycle of anxiety. "As long as we deal with our emotional problems at only an abstract conceptual level," they write, "change in these problems is unlikely to occur."

When we're anxious, we often feel an urgent pull to "figure out" our problems or wonder "what if" some catastrophic scenario came to pass, yet worry is a psychological dead end. Recall what we learned from LeDoux's fear research: fear springs from the subconscious amygdala, not the thinking cortex, so trying to think your way out of feeling afraid is like using a hammer to twist in a screw. Our fears are not cognitive puzzles to solve. Seductive as it can be, worrying gets you nowhere.*

Mistake #4: Suppressing your feelings and thoughts

One day, when Leo Tolstoy was a young boy, his older brother gave him a simple challenge: Go sit in the corner, and don't come out until you have no thoughts of white bears. Whatever you do, the brother said, just don't think of a white bear. The task appeared easy enough, but as the legend goes, little Tolstoy logged hours in that corner trying to block the sudden gush of white bears now flooding his thoughts. The harder he tried to suppress thoughts of bears, the more of them he saw, battalions of them marching in

---

*The logical question to ask here is "Okay, then how do I curb my worry habit?" We'll discuss this, along with a cornucopia of other concrete ideas for handling fear, anxiety, and stress, in the Conclusion.

lockstep through his mind. Finally, he gave up; as he probably should have guessed from the sadistic gleam in his brother's eye, the mission was impossible. It was a mental trap.

This might seem like a trivial thought experiment, but the Harvard psychologist Daniel Wegner has spent two decades teasing out its implications for our mental health—namely, that our attempts to control our thoughts and emotions tend to backfire. In one early experiment, Wegner told two sets of subjects to ring a bell every time they thought of a white bear over a five-minute span, with one group asked *never* to think of white bears and the other group asked *only* to think of white bears. Paradoxically, the ones told to suppress white bear thoughts ended up ringing the bell *more often* than those told to imagine them from the start. This effect extends to our emotions as well. When Wegner placed students under mild stress and then instructed them to "relax"— in effect, to suppress their feeling of tension—they grew more anxious than before. The more "motivated" they were to kill their feeling of anxiety, the harder it was to do.* As the Columbia University psychologist Kevin Ochsner explains, emotional suppression boomerangs on us. "Trying to hide your feelings is distracting, it worsens memory, and it boosts some measures of nervous system activation," he said. In fact, psychological data from both world wars showed that British troops, with their famous "stiff upper lip," were more prone to subsequent mental distress than their counterparts from emotionally open countries like France.

Wegner's studies draw out the important distinction between stifling difficult feelings and simply letting them be. Just as it's

---

*This isn't to say that trying to relax is *impossible*, just that it's easier to do it when you focus on something other than your emotional state—like your slow, placid breathing.

pointless to try to worry your way out of anxiety, it's futile to try to overcome a fear by bottling it up or ceasing all anxious thought. Instead, today's psychologists teach people to experience their fears openly, and to note their worries and negative thoughts without getting entangled. Which is an especially tricky skill to master, because of…

Mistake #5: Buying into distorted thinking

For his 2008 book *Click*, the online-behavior researcher Bill Tancer sifted through millions of search engine queries to compile a list of more than a thousand unique fears, from big ones like spiders down to esoteric aversions like "fear of capital letters." The top ten most-searched fears he found were, in descending order, flying, heights, clowns, intimacy, death, rejection, people, snakes, success, and driving. It's not a scientific survey, of course (most studies peg public speaking as our most common fear), but the results reveal something interesting. We believe these things to be grave threats, yet with the obvious exception of death, none of them is likely to cause us serious harm. As we learned in our exploration of the amygdala's inner world, our fears don't always obey reason; if they did, many psychologists point out, we'd all list legitimate evils like guns and heart disease as our most urgent concerns. But instead, we engage in a maelstrom of twisted thinking about these idiosyncratic, exaggerated fears.

I often see a bumper sticker around town that cautions, "Don't believe everything you think," and this is especially true with anxious thoughts. Our fears and worries are very convincing, but they're fraught with biases and inaccuracies. Said Craske, the anxiety expert, "When we're treating fears and phobias, the cognitive element of the process is really about interrupting things like the tendency to overestimate the likelihood of negative events, or to catastrophize the meaning of things." The more anxious you are about something, Craske explains, the more error prone your predictions about it become. In the immediate aftermath of 9/11, for instance, Americans told researchers that the likelihood of the average citizen getting injured by a terrorist attack in the next year was 48 percent. (As it turned out, it was 0 percent.) Fear-tinted thoughts also veer toward needlessly catastrophic conclusions: your friend didn't say hello to you on the street because she hates you, not because she simply didn't see you. Likewise, the cognitive therapist Robert Leahy points out that anxious thoughts tend to begin with phrases like "It would be awful if..." or "I couldn't handle it if...," even though research shows that people cope with negative outcomes far better than they anticipate.

Naturally, not all fearful thoughts are biased. Some things really are dangerous. But believing every worry without throwing at least one industrial-strength grain of salt into the mix only fuels our fears.

Mistake #6: Living a modern life

A century ago, the celebrated psychologist William James noted that modernity had insulated us so well from mortal dangers that "in civilized life...it has at last become possible for large numbers of people to pass from the cradle to the grave without ever having had a pang of genuine fear." So James may have been surprised to learn that the more "civilized" we become, the higher our anxiety soars. In survey after survey, developing-world countries show only a fraction of the anxiety issues that modernized nations face; the 2002 World Mental Health Survey found that Americans were nearly five times more likely than Nigerians to experience clinically significant levels of anxiety in any given year. And we're still on an upward slope. In her book *Generation Me*, the psychologist Jean Twenge describes the staggering uptick in anxiety her research has found among young people: "Anxiety increased so much that the average college student in the 1990s was more anxious than 85% of students in the 1950s and 71% of students in the 1970s." Between 1957 and 1996, Twenge writes, Americans also became 40 percent more likely to describe themselves as having been on the verge of a nervous breakdown over the past year.

Psychologists have a variety of opinions about why modern living seems to steer us toward ever-growing anxiety. One proposed culprit is information overload. Today's average Sunday newspaper includes more raw information than people in earlier eras would take in over the course of a few *years*, and this avalanche of data is increasingly of the alarmist, amygdala-inflaming variety. If a TV newscast isn't covering a grisly double homicide, the anchor is teasing a story about the hidden threat *in your own home*. "The media does this to us," says Evelyn Behar. "It's always reporting that this thing causes cancer or that thing can kill you. We live in a culture where fear is used to motivate us." The fear researcher Michael Davis, on the other hand, points to modern society's lack of close community and family connection, as people migrate across the country to insular suburbs. "If you've lost the extended family and lost the sense of community, you're going to have fewer people you can depend on, and therefore you'll be more anxious," he told me. Yet another theory is that we've fallen victim to "feel-goodism," the idea that bad feelings ought to be annihilated, controlled, or erased by a pill, which puts us at loggerheads with the reality of the human condition—sometimes, things are just tough. Regardless of which theory you believe (and it could be all of them), the overarching point is this: modern habits and anxiety often go hand in hand.

Mistake #7: Disappearing into the future

As animals go, humans are almost embarrassingly scrawny and frail. We have no fur or hide to protect us from the weather, we have no sharp claws or teeth to defend ourselves, and we're far weaker than our larger primate cousins. In other words, we are the nerds of the animal kingdom. Our main advantage over other species is our massive prefrontal cortex, which enables us to design tools, communicate through language, and form complex plans for the future. Only we can anticipate and thoroughly prepare for potential threats. But as Joe LeDoux explains, this cognitive benefit has its drawbacks. "Bigger brains allow better plans," he writes, "but for these you pay in the currency of anxiety." After all, anxiety is all about the future—about worrying over hypotheticals and figuring out what-ifs. Sometimes planning serves us well, and sometimes not. Our anxieties can make us withdraw into an imagined future filled with threats that don't yet exist and likely never *will* exist. We forget about right now.

In our post–New Age world, where the hazy admonishments of the 1960s now appear in credit card commercials, the advice to "Be here now" has become something of a cliché. Still, when it comes to fear and anxiety, the recommendation holds true. As we'll see later, if you are thinking anxiously, then you are by definition living outside the present moment—and, in a sense, even ignoring reality itself.

Mistake #8: Avoiding what scares us

Finally, we come to the single most important error we commit in dealing with fear, a blunder so egregious that it needs to be set in italics: *avoiding the situations that make us anxious*. On the surface, avoidance actually makes perfect sense; why *wouldn't* we shy from things that seem like they could hurt us? But the hidden negative effects of this habit can be crippling. If fear is like a living organism in the mind, avoidance is its primary means of self-preservation. Without exposing ourselves to the things that trigger our fears, we never get a chance to learn that we can cope, or that our catastrophic worries are wrong, or that the things we fret about really aren't going to tear us limb from limb. Avoidance ensures that the fear lives on.

This link between evasion and fear preservation is so strong that psychologists consider chronic avoidance to be the central feature of an anxiety disorder, a state of anxiety persistent and intense enough to interfere with a person's life. "Avoidance is a factor in all anxiety disorders—it just takes different forms," Michelle Craske said. "So in obsessive-compulsive disorder, the avoidance behavior is doing the compulsion. Or in straight-out phobia, the avoidance behavior is to refuse to approach the stimulus." Therapy and medication work wonders to treat anxiety disorders, but many with severe anxieties prefer to plan their lives around avoiding their fears instead of going through the difficult process of dealing with them. One famous example of this is the football broadcaster John Madden, who nurses a legendary fear of flying. After Madden panicked from claustrophobia on a flight in 1979, he spent the next thirty years crisscrossing America for his announcing gigs in a bus, the "Madden Cruiser." Flying could have saved him thousands of travel hours, yet Madden refused to entertain the idea of reboarding a plane.

Because avoidance is such a disastrously counterproductive strategy, you might be expecting me to proclaim here that we should "face our fears" or "confront" them, yet these ideas are problematic as well: they suggest that fear is an adversary, that working through anxiety is a battle with a winner and a loser. This attitude, as we'll soon see, couldn't be more wrong. Fear isn't something you avoid, fight, or figure out. It's something you embrace and learn to work with. So, if you'll permit me to swap one hackneyed cliché for another, a far better strategy is to open up to our fright and "go with the flow." Fear and anxiety are a great, rushing river upon which we float in our bobbing little kayaks. We can paddle furiously against the stream in a futile struggle to get upriver and avoid the rapids, or we can work with the current and use our energy to navigate the challenges ahead. The choice is always ours.

Extinction-Level Event

The advice to expose yourself to your fears is far from brand-new, of course. In the nineteenth century, for example, the fretful Danish philosopher Søren Kierkegaard wrote, "We cannot mature and be fully creative by burying or displacing anxiety, but only by moving through it." We've seldom been very good at following this prescription, perhaps because it has historically lacked scientific punch; no one ever tells you *why* you need to experience your fears, just that you should.

Enter our friends the neuroscientists, and the mental protocol they've dubbed "fear extinction." On the day I spoke with Michael Fanselow, the UCLA fear researcher we first met in the last chapter, he had just weathered an experience that underlined the weirdness of this extinction process. Fanselow, you'll recall, was a pioneer of fear conditioning studies—rat hears tone, rat gets shock, rat learns to fear tone—and earlier that day, his amygdala had acted out in a way that he'd often seen in his rodent subjects. Back in 1994 Fanselow lived through the devastating Northridge earthquake, which killed fifty-seven people, leveled buildings and highway overpasses across Los Angeles, and left millions—Fanselow included—with a bone-chilling fear of future tremors. "Whenever I'd feel a vibration after that, I'd have a pretty strong fear response," he told me. "When I'd be standing in a parking garage and I'd feel the shaking from cars going by on other levels, that really sent me into it." His intimate knowledge of the amygdala provided little relief. "I could say, 'Hey, I just made this cool species-specific defense response,'" Fanselow recalled, "but recognizing that I had some good fear conditioning didn't really matter. What mattered was getting exposed to that fear enough times so that it went away." In the broadest possible sense, this is how our fears go "extinct"; with repeated exposure to what scares us, the brain learns not to be afraid anymore.

But as Fanselow's experience earlier that day showed, things aren't *quite* that simple. His earthquake fear level had hovered near zero for years, and then a moderate tremor jolted L.A. a few hours before we talked, rekindling a long-dormant dread in his amygdala. "The earthquake definitely managed to get my attention," he reported with a laugh. Fanselow had just experienced the same psychological phenomenon that his lab rats frequently faced, the "spontaneous recovery" of a fear. Even though his fear of quakes had diminished, he explained, the original fear memory never truly disappeared—it had just been covered up. So in the same way that his deconditioned lab rats will often show a fresh burst of fear to that digital tone if sufficient time passes, Fanselow's fear of earthquakes spontaneously recovered; the old fear slipped through a mental crevice and expressed itself anew. This is the bad news: as we saw in the first chapter, our fear memories are pretty much permanent. But our brains also feature a handy neural mechanism that, with a few hitches, allows us to get over virtually any fear we might harbor. Because of the fear extinction process, none of our fears is an immovable object—which is why

Fanselow wasn't too concerned about the mental aftershocks of that day's quake. "My amygdala's firing a lot initially," he told me, "but soon enough it'll see this isn't something to worry about."

Here's how fear extinction works its magic. The extinction process consists of two key ingredients: exposure to whatever it is that frightens you, and, surprisingly enough, *fear itself*—the more the better. We already know that the amygdala has a bit of a paranoid, unruly streak, often setting off unwarranted fear reactions to vague potential threats before the conscious mind can figure out the full story. But fortunately, the amygdala isn't a moron; it's just a slow learner when it comes to seeing when *not* to be afraid. To illustrate this, let's now play out the familiar fear-triggering process a bit further. Imagine that you have a snake phobia (or ophidophobia, for those keeping track at home), and as you're leafing through a nature book, you suddenly land on a photo of a coiled sidewinder sunning itself on the desert sand. As we've seen, this will displease your amygdala; with its fuzzy vision, it won't be able to tell that the snake picture isn't a real threat. You will feel afraid. Your impulse will be to close the book and make the terrifying image go away. But what if you didn't do that? What would happen if you stared at that picture without turning away, and just let yourself feel scared?

This is what would happen: you would be afraid for a while, and then at some point, your subconscious mind would begin to piece together an important truth. "Wait a minute," your brain would say. "Nothing's happening. This snake photo isn't attacking me." If you looked at that photo once every day, it would lose a shred of emotional impact each time—and were you hooked up to a brain scanner, you'd see a region called the ventromedial prefrontal cortex flexing its cognitive muscle more and more, gating off the amygdala's ability to fire off a fear reaction. "Through experience, the prefrontal cortex is learning to say, 'Okay, amygdala—in this situation, shut up,'" explained Fanselow.

Evolution, it seems, wanted us to learn our fear lessons fast in order to maximize survival, but it also didn't want us to miss out on valuable opportunities if that fear turned out to be misplaced. So, if you're exposed to a fear-provoking stimulus for enough time and nothing bad happens, your higher brain learns to shut down the amygdala's flare-ups before they even start. We don't need to "solve" anything intellectually. The learning just happens automatically, if we let it. "When a person has a fear, it's not a case where you need to figure something out rationally," said Fanselow. "That isn't what you need. What you need is exposure." In a sense, then, being afraid of something doesn't necessarily constitute proof that the thing is inherently bad; it may just mean you haven't spent enough time hanging around it. These new cortical lessons might fade over time, as in Fanselow's spontaneous recovery of his earthquake fear, but you can relearn them more quickly and easily than before, like taking a refresher course in school.

Neuroscientists are still a bit murky on how the amygdala and the ventromedial prefrontal cortex work together to hit the mute button on a fear, yet for us, the main message is all that's important: to get over a fear, you have to expose yourself to it, and you have to feel afraid. Naturally, this is easier said than done—our brains don't exactly surge with joy at the prospect of stewing in dread—but moving *through* fear is the only way *out* of it. When I asked Fanselow if it's possible to get over a fear without exposure, he replied, "There's no evidence that that will ever happen. Even time doesn't seem to heal fear per se. What time does is give you the opportunity to expose yourself to it, even if it's not in full doses." In fact, clinicians who treat phobics aim to make their patients feel legitimately afraid during therapy, because this gives their brains the best shot to learn that elevators or bridges aren't really going to hurt them. "We call it peak and pass," said Evelyn Behar, the University of Illinois psychologist. "You want people to peak in anxiety, so you get as high a level of physiological response in the face of a threat as possible, and then you allow that to pass naturally. If you don't run away from it, the response will go down. But without that fear response, you will never get anywhere with anxiety."

The most extreme example of this meet-it-and-feel-it approach to fear comes from the mind of Boston University's David Barlow. At his Center for Anxiety and Related Disorders, Barlow practices a treatment method called interoceptive exposure, which has one solitary goal: scare the hell out of the patient, to teach him he can handle it. If standard talk therapy is a nice stroll through the countryside, Barlow's method is a headlong sprint along the edge of a cliff. Patients who fear public speaking don't learn how to relax in front of a crowd; instead, Barlow jacks up their nerves with copious coffee and then sends them out in front of an audience of fellow patients whom he has instructed to act as boorishly as possible. He locks claustrophobics in the trunks of cars, sends acrophobics up high, and generally teaches patients to unflinchingly welcome whatever scares them, until either the terror goes away or they get so used to the fear that it no longer bothers them—they've become completely okay with their anxiety. "A good patient doesn't only accept fear, he courts it, chases it," Barlow told the *New York Times*. More conventional therapies attempt to help patients relax, but Barlow contends that focusing on becoming calm can send a false message that fear is dangerous. "There's a place for relaxation," explained Craske, a Barlow collaborator, "but if the crux of the problem is that a person is afraid of feeling fear, then too much focus on relaxation simply feeds that fear." Interoceptive exposure strikes many people as mad, but the method is as effective as it is extreme. Barlow says his technique helps 85 percent of patients, often in a week or less.

The primary obstacle with an approach like Barlow's is getting people to give it a shot. (A sales pitch of "We're going to have you jump directly into the abyss of your fiercest terror" doesn't get people racing through the door.) Exposure is tough. The alternative, for many, is to rely solely on medication, which is as easy as swallowing a pill. This is why sales of antianxiety drugs have rocketed into the stratosphere of late: it seems like a trouble-free fix. These drugs are very effective at curbing the brain's ability to produce fear, especially the class of medications called benzodiazepines, which includes Valium and Xanax. Benzodiazepines are useful tools, but they present a serious problem. If they keep you from feeling fear altogether, you're never going to get over what scares you. Some psychologists refer to these drugs as "mental Teflon"; because the fear can't express itself, new learning just bounces off. Nowadays, psychiatrists and doctors favor antidepressants in conjunction with therapy for anxiety issues because they help make people more positive and less avoidant, putting them in a better position to meet their fears.

But while no pill can solve our issues with anxiety on its own, the Emory University neuroscientist Michael Davis has found a drug that may be the next best thing. In the early 1990s, Davis discovered that a protein called N-methyl-D-aspartate, or NMDA, plays a critical role in helping the amygdala learn when not to be afraid during the extinction process. "NMDA is a biochemical coincidence detector," Davis told me. "It's built to associate two things, and that's what learning is all about." Davis reasoned that if he could tweak our NMDA receptor sites to make the brain learn more efficiently, then he could speed up fear extinction. To do this, he used the out-of-fashion antibiotic d-cycloserine, which had long been used to treat tuberculosis. Somewhat unexpectedly, low doses of d-cycloserine were found to work as a sort of cognitive superdrug, juicing up the brain's NMDA sites and greatly accelerating learning. When Davis combined a dose of d-cyclo-serine with exposure treatment for phobias, the initial results were staggering. In one study, Davis took twenty-eight severely acrophobic volunteers and gave half of them the drug and half of them a placebo. Then he put them through a harrowing virtual reality simulation of a ride in a glass elevator. The ones who took the drug showed the same reduction in fear after two virtual real-ity sessions that the placebo patients showed after eight sessions; the drug made fear extinction work four times as fast. Despite these electrifying results, however, Davis has had trouble raising the millions needed to test and market the drug; since d-cycloser-ine is already available generically, investors worry about its potential profitability.

So for now, those seeking to overcome a fear are left with the best God-given tool available: exposure. But although exposure of some kind, even just picturing what you're afraid of, is always nec-essary for fear extinction, it's not always sufficient on its own; if we don't adapt our *attitude* as well, we can still sabotage the process. Explains Craske, a snake phobic can bravely face down a harmless garter snake, knuckles white from strain, worries and negative thoughts cascading through her head, and not progress along the road to extinction. "Some people don't learn from the experience," she said. "I've seen it many times. They're holding on for dear life to get through it, telling themselves they can't take this, and no new learning happens." Our mind-set matters greatly in handling fear wisely, but the thinking brain's role isn't to figure out anxiety or analyze it away. Instead, the mind's job in fear extinction is sim-ple: be open to whatever comes up, and get out of the way. Anxiety can lead us down a number of unproductive roads—worrying, believing distorted thoughts, leaving the present moment—yet the best cognitive approach to fear and anxiety can be summed up in a single word. It's called mindfulness.

Naked Attention

Somewhere around 2,500 years ago in the foothills of the Himalayas, a twenty-nine-year-old prince named Siddhartha Gautama snuck out of his family's palace in the middle of the night and set off on a quest to find himself. Gautama had spent his entire life in the ancient Indian equivalent of a gated suburban community, and the young prince had grown deathly tired of the sheltered world his father had provided. If Gautama were to visit a psychiatrist's office today, he'd probably walk out with a diagnosis of depression and a prescription for Prozac, but since no such drug existed then, he chose another route out of his spiritual quagmire: intensive meditation. For years, he apprenticed himself to various wise men. He learned deep yogic meditative practices, and he tried to sort out the universe intellectually. He spent months in the forest attempting to gain control over his mind while fending off a feeling of horror (tiger attacks were always an issue), and he ventured so far into ascetic self-deprivation that he almost died of starvation. To his frustration, nothing seemed to liberate him from his psychic suffering.

One day, as Gautama was recuperating from his bout with malnourishment, he decided to change his tactics. Instead of striving for an intellectual solution to his problems or struggling for power over his feelings, he did something very simple: he just paid bare, wordless attention to everything that happened to him in the present moment. Walking down a path, chewing his food, breathing in and out, Gautama became quietly conscious of how the experience felt, neither clinging to it nor pushing it away. When an emotion arose, he watched it swell and fade. When thoughts sprang into his head, he noted them and then let them dissolve. Gautama later said that this made him realize his mind was like a monkey swinging frantically through the trees—"it grabs one branch, and then, letting that go, seizes another"—and that as he gained some distance from this monkey mind, he began to "see things as they really are." He called this new state of non-judgmental awareness *sati*, or mindfulness, and it formed the cornerstone of his life. At the moment he attained enlightenment, Gautama became known as the Buddha, and he soon began spreading the word about *sati*. In one famous early discourse, the Buddha said of mindfulness, "This is the one and only way, monks, for the purification of beings, for the overcoming of sorrow and lamentation, for the extinguishing of suffering and grief, for walking on the path of truth, for the realization of Nirvana."

Don't worry: we're not about to lapse into a long soliloquy about Eastern philosophy. There's a very important point for us to consider in the Buddha's words, however, even if we don't subscribe to the whole karma/reincarnation side of Buddhism. When it comes to dealing with emotions like fear, it turns out that the Buddha was several thousand years ahead of his time. Recent research on mindfulness practices has sent psychologists racing to preach a mindful approach to anxiety, worry, and all breeds of mental strife. We all know that the Buddha was a calm customer, but in his subtle understanding of the workings of the amygdala, it seems he was a gifted neuropsychologist as well.

In his book *Mindfulness in Plain English*, the Sri Lankan Buddhist monk Bhante Henepola Gunaratana describes mindfulness as "wordless bare attention" to what's happening right now. "It is not thinking," he writes. "It does not get involved with thoughts or concepts. It does not get hung up on ideas or opinions or memories. It just looks." Since mindfulness is by definition a wordless, participatory activity, we could spend fifty pages talking about it and get nowhere; it's something you learn by doing. But for our purposes, the vital idea to take away from Buddhist mindfulness teachings is that the concept encompasses precisely the opposite attitude that we saw in our eight anxious mistakes.* Westerners often think of Buddhists as fleeing reality to lounge in meditative oblivion, but the truth is that the Buddha saw himself as a man trying to experience the world as fully as possible. Frequently, those who were awestruck by Gautama's tranquility asked him if he was some sort of god or magician. "No," he replied, defining himself instead in three simple words: "I am awake." For him, mindfulness was about openness and awareness, not control, and he advised his monks to thoroughly feel their emotions without getting in the way. The Buddha cautioned against brooding over worries and getting entangled with distorted thinking, instructing his disciples to learn to stay anchored in the present moment and to take note of each anxious thought as one would a passing cloud, letting it float by without grasping on to it.

In other words, the Buddha was a master amygdala whisperer— he had learned to work with his brain's fear center rather than against it. It's hardly a surprise, then, that new research has provided a huge boost to mindfulness's reputation in psychological circles. Take the common Buddhist practice of "noting," for example: practitioners learn to label their worries and feelings with a simple tag like "thinking" or "anger," taking note of them mindfully without engaging them directly. In a 2007 study, the UCLA psychologist Matthew Lieberman showed thirty volunteers fear-provoking images and then asked them to note their feelings ("I feel afraid") as he monitored their brain activity. Upon seeing the unpleasant images, the subjects' amygdalae lit up at first, but the labeling process soon sparked activity in the right ventrolateral prefrontal cortex, damping activity in the amygdala.

---

*Except for being born that way, of course—not even the Buddha could do anything about that one.

Lieberman believes this mindful noting—the simple act of putting our feelings into words—helps the brain disambiguate our emotions and provide a level of detachment from them. "One of the ways labeling is useful is in talking with other people," he told me. "If you can get someone to talk about their feelings, it'll end up being beneficial to them in ways they may not realize." (Writing about how we feel in a journal serves the same purpose; it helps us sort out emotions, like anxiety, on a deeper subconscious level.) In a similar vein, breakthrough mindfulness studies by researchers such as the University of Wisconsin–Madison's Richard Davidson—who has cataloged a number of ways that practicing mindfulness remodels the brain itself—have provided such encouraging results in dealing with anxiety that a well-funded research boom is under way as we speak, with more than a hundred papers being published on the topic each year.

Perhaps the most promising new mindfulness-based approach to anxiety comes from a man who made every mistake with fear one can make, only to emerge from his ordeal wiser, and with an innovative approach to mental life. One day in 1978, when he was a young assistant professor at the University of North Carolina at Greensboro, a stressed-out Steven Hayes suddenly grew so overwhelmed with anxiety during a psychology department meeting that he could barely speak. He was having his first panic attack. The spooked Hayes soon launched into a frenzy of counterproductive behaviors to try to gain control over his anxiety. He squirmed out of his now horrifying class lectures as much as possible, and he began avoiding anything that might make him feel trapped and trigger his anxiety: elevators, restaurants, planes, even phone calls. The more he fought his anxiety, the worse it got. "The only thing that mattered was getting the anxiety to go away," said Hayes, who now teaches at the University of Nevada,

Reno. "Every moment of your waking life becomes about anxiety. Well, try to live that life, see how that works. It just sucks."

Hayes was fighting the fire of his anxiety by spraying it with gasoline—avoiding, worrying, seeking unattainable control over his feelings. "Our normal mode of mind, when applied to fear, makes us do all the wrong things," he told me. "You grab that fly-paper monster called anxiety and struggle with it, and it'll wrap itself completely around you. You'll be stuck so tight, you can't move." It took several years before he realized, in a flash of insight, that his adversarial approach was only exacerbating his dilemma. "If I had to pick a single moment, it was in the middle of a panic attack, and I almost shouted it out loud, like in a Tarzan movie," Hayes recalled. "I yelled to myself, *'You can make me afraid, you can make me hurt, but you can't make me turn away from my own experience.'*" He realized that the only way to stop this downward spiral was to become mindful, to attempt to coexist with his fear instead of pushing it away. "I quivered and shook and had panic attacks for years after that," he said, "but I was on a different path, and it was one that was opening up and positive."

Now, Hayes has been panic-free for decades, and he's channeled his experience into what he calls acceptance and commitment therapy, a mindfulness-based therapeutic technique that has shown incredible results with anxiety of all kinds. At the center of ACT lies an essential truth that we'll reflect back on many times as we look at the specifics of dealing coolly with fearsome situations: *fear is not the enemy*. Unpleasant as it sometimes feels, fear has no sinister intentions, nor does it carry hazardous side effects. Fear and anxiety are simple physiological and psychological reflexes, like the sensation of hunger or pain—impulses that help us survive, even if they crop up at inopportune times. Said Hayes, "It kind of dumbs you down to treat fear the way it's normally treated, where you're not open to what's inside fear that might be useful."

Hayes's ACT model takes a hands-off stance toward the things we can't change—automatic thoughts, emotions, past events—and focuses instead on helping people accept their feelings, mindfully disentangle from worries, and move on with what they really care about in life. Indeed, psychologists generally agree that changing behavior, not thoughts or feelings, is the best route to improved mental health. Explained Behar, "Behavior always outweighs cognition. If you can make yourself behave in a healthier way, it can actually change your thoughts. But you can sit around forever waiting for your thoughts to change, and it'll never happen." As a child of the sixties, Hayes saw the futility of trying to make Westerners buy into Eastern thought, and he remains adamant that one needn't meditate, join an ashram, or wear loose-fitting clothing to reap the benefits of the mindful life. When people live full lives and openly accept their thoughts and feelings, he believes, they combine the neurological power of the fear extinction process with the equanimity of mindfulness. If you feel afraid, that's great: you're exposing yourself to the emotion, and the more you let it play out, the more you allow your brain to process it. If your mind shoots out anxious worries, that's fine, too: it gives you a chance to practice being mindful of the monkey mind's harmless chattering, without getting bound up in never-ending internal debates.

As your attitude toward the world becomes more open and accepting, Hayes believes, fear no longer seems like an adversary. "When you're not standing in relation to fear as some sort of psychologically horrible event that has to be run from, you start getting this new flexibility," Hayes said. "You learn how to feel anxious, period, end of story, instead of fighting it." So far, the research has agreed with him: in several studies, ACT has outperformed older cognitive therapy models (which tend to involve challenging your anxious thoughts directly) in treating anxiety and phobias. In the January 2006 issue of *Behaviour Research and Therapy*, Hayes summarized thirteen clinical studies that pitted ACT against older therapeutic methods, and in twelve of those cases, ACT outperformed all comers in reducing psychological distress.

Some have accused mindfulness boosters of presenting the practice as a sort of miracle psychic cure-all, and the criticism is a fair one. There's still much solid research left to be done on the subject, and the irrational exuberance of some mindfulness advocates has the whiff of psychological faddishness. None of this would have surprised the Buddha himself, who went out of his way to point out that mindfulness isn't a secret lever one flips, thereby rendering everything in life suddenly hunky-dory. Mindfulness, he said, is a practice—a skill one hones slowly over time. "Just as the ocean slopes gradually, falls away gradually, and shelves gradually with no sudden incline, so in this method, training, discipline, and practice take effect by slow degrees, with no sudden perception of the ultimate truth," he told his monks. Mindfulness involves steady effort. Results take time and patience.

But the Buddha also liked to point out something he considered equally important: right at this moment, each of us already has everything we need to begin transforming our relationship with fear for the better. All it takes is openness, determination, and a willingness to feel afraid.

Leaning into It

Danny Forster may be the only television personality in history whose show has featured his fierce, real-life phobia as a main character. This wasn't quite how he planned things. Since 2006, Forster has hosted the Discovery Channel program *Build It Bigger*, which is a show about…well, building big things. The job requires Forster, a trained architect in his early thirties, to participate in the construction of giant skyscrapers and lofty bridges as he explains how these towering behemoths come into existence, which would be well and good if not for one tiny problem: Forster has a severe fear of heights. One might think this would disqualify him from getting the post in the first place, but when he answered the producers' Craigslist ad for an architecture show host, Forster—still in graduate school and filled with a boundless enthusiasm for building design—had no clue how serious his phobia really was. "I applied for this job because I'm passionate about buildings," Forster told me. "I didn't necessarily want to get on *top* of the buildings."

It was on his very first scouting trip as host that Forster's true acrophobia reared its head. He and the show's producers had flown to Glendale, Arizona, to peek in on the construction of the show's first target: University of Phoenix Stadium, the new $455 million, state-of-the-art home of the NFL's Arizona Cardinals. They planned to have Forster spend a lot of time working on the structure's 240-foot-tall fabric-lined roof, but when it came time for Forster to climb a ladder to the top, he balked. "Right then, I knew there was going to be a problem," he recalled. "For the two weeks before the actual shoot, there was this flurry of e-mails at the Discovery Channel saying, 'Holy shit, did we hire the wrong guy?'"

If Forster had harbored any hopes of seeing his fear of heights pass quickly, a nightmare experience during the actual Arizona filming shattered them. On his third day of shooting, Forster, his cameraman, and a heavyset veteran ironworker rode a couple hundred feet up the side of the stadium in a manlift—a mobile, telescoping aerial elevator with an open basket on top—to hang iron beams in the structure's upper framework. As they started to come back down, the manlift suddenly locked up and issued an ominous beeping noise. The basket began a stomach-churning sideways tilt. "When this old-school ironworker turned the thing off and took a deep breath, I was like, *Fuck*," Forster said. "I weigh about a hundred and sixty pounds. My camera guy weighs about one fifty, plus the camera equipment, and the ironworker was about three hundred. And the weight limit for the lift was five hundred pounds. It was so bad that the back wheels of this thing on the ground were actually tipping up." The lift eventually started working again, and the televised show only hints at the danger they were in, but as Forster told me, "We legitimately almost died. When we pulled the ironworker's mike after it was all said and done, he said, 'Thirty-eight years hanging iron and I've never come that close to getting killed.'"

Forster's fear only intensified after this near miss, but as his commitment to sharing his love of architecture propelled him on through new episodes, his Discovery Channel bosses quickly saw that his acrophobia made the show more human. Forster no longer saw the point in trying to hide his fear—"If I tried to fake it and look brave, it'd just come off as a show about a guy who was constipated or something," he said—and his visceral reaction to hanging off the sides of buildings brought home the scale of the colossal structures. His acrophobia, he found, was becoming a *selling point* for the show. Soon, almost every *Build It Bigger* episode featured shots of a haggard-looking Forster contemplating a soaring perch with pallid dread, with a few common fearful phrases thrown in: for example, "This is not a good situation right now," "Am I getting pale?," and *"Oh oh oh oh!"* The toughest experience yet for Forster came in Shanghai, which the show's crew visited to shoot the new 1,600-foot-tall Shanghai World Financial Centre, the world's second-loftiest skyscraper. In China, buildings fly up fast, unencumbered by pesky obstacles like worker safety regulations, and the chaos of the Shanghai tower nearly had

Forster in hysterics. First there was the thousand-foot ride up in a makeshift elevator—barely more than a steel cage—that was bolted to the side of the building. Then there was the transition from the cage of doom to the unfinished patchwork floor in the sky, with only a few planks and beams standing between the workers and the ground below.

Worst of all, though, was the view straight down. "Here's a funny story about this," Forster recalled. "I usually use my cameraman like a barometer: if he's relatively calm, everything's probably okay. The elevator core of this building in Shanghai is basically like a big concrete sleeve, a vertical empty cylinder so wide that it fits ninety-seven elevators. Now imagine walking to the edge of that ledge and looking down sixteen hundred feet into this concrete death hole. It's like the originating spot on earth was down that tunnel. So I said to my cameraman, 'Jason, go take a look, it's crazy,' and he walks over with the camera on his shoulder, leans over the edge to look down the hole, steps away, and vomits. So I was like, 'Okay! This really is horrible.'"

But as harrowing as these experiences were for Forster, he began noticing slight changes in how he felt as he kept plugging along with what he loved: communicating his excitement about buildings to a national audience. His hosting duties forced him to behave flexibly even as his terror surged; he made himself devise intelligent patter and keep at his construction work despite his urge to flee to lower ground. His willingness to expose himself to frightening situations showed his amygdala that he could handle himself in midair, and it began firing less and less. "You know, it's always been pretty difficult for me to do this stuff, but I always do it," Forster said. "I've never said, 'We can't do that.'" His openness about feeling afraid eased the tension of dicey situations, and even ended up endearing him to his burly construction worker colleagues. At first, Forster had been an inveterate worrier: "I could write a book with the internal monologue," he told me. "I would scenario out all the different ways I could die. I could tell you about the hungover laborer in the steel mill who didn't roll out the steel properly that Wednesday so when it got delivered, it had a little failure in it that would just happen to be where I'd be standing." Over time, however, he learned to step off these anxious trains of thought and watch them pass by. "You kind of let them wash over you and dissipate and don't judge them, just be in the moment," he said. "And what I'm finding now — and it's a very new experience for me — is that I can enjoy it just a little. I don't want to give myself too much credit, but I'm just beginning to feel that twinge of exhilaration where I *like* heights."

Without really intending to, Forster put into practice much of what we've discussed in this chapter about the right way to be afraid. He exposed himself to the situations that terrified him, opened himself up to feeling scared, stopped trying to figure everything out, and just got on with what he valued in life, fear or no fear. It's a basic technique that any of us can apply, even the seasoned hypochondriacs we met at the beginning of this chapter. If the average hypochondriac's main problem is that he can't accept the possibility that he might be ill, then a skilled psychologist can help him learn to live with uncertainty, live with the reality that he can't control everything, and just *live*. Arthur Barsky, the hypochondria expert, says that having severely health-anxious people stop their counterproductive avoidant behaviors — checking symptoms, surfing health websites, scheduling unneeded tests — and return to living full lives often yields results that surprise long-suffering hypochondriacs. "Ultimately, what happens for people who get better is not that their physical symptoms or anxiety completely goes away," Barsky told me. "It's just that it becomes much less intrusive, less bothersome. It doesn't run their lives. Someone will tell me, 'You know, if I stop to think about it,

I still have that lump in my throat, but I just don't think about it that much.' It doesn't interfere with their quality of life." Their anxiety doesn't need to be such a bad thing after all.

From this point on, as we explore how people handle a variety of fearsome situations with coolness and poise, this will be a common theme. Heroic soldiers, clutch athletes, unflappable pilots, and steady-handed doctors all feel plenty of fear—they just don't see it as something awful. Instead of fighting or avoiding their fears, they tend to concentrate on doing what has to be done. Their relationship with fear isn't adversarial, but accepting. This open approach isn't always easy, and it doesn't provide an instant cure-all, but it's the only road out of the woods.

As Danny Forster learned while swaying several hundred feet up in the air, fear is something he had to learn to lean into. "When I'm with the ironworkers, they always say that once they've tied off their ropes and once their harness is set, they like to lean back into it," Forster said. "Rather than cling to the column, they like to lean right through the harness so the tie-offs are being put under tension. I always used to ask, 'Why would you do that? What if the carabiner breaks?' And they'd always say, 'You've got to trust the equipment.' It reflects back to them that everything's working. So I've started to do that a bit. You tie off. You lean back into it. You don't fall. Okay."

CHAPTER THREE

The Zen of Shock Trauma: Stress, Strain, and Coping with Chaos

I f you think your job is stressful, meet Dr. Thomas Scalea.

A compact and confident man in his late fifties, Scalea is the physician in chief of the R. Adams Cowley Shock Trauma Center at the University of Maryland Medical Center in downtown Baltimore. When very bad things happen to people in the Baltimore area—car wrecks, stabbings, freak accidents, and everything else that suddenly thrusts a person onto the precipice of death—paramedics rush them directly to Shock Trauma, the best unit of its kind in the nation, where Scalea and his team specialize in working under pressure to outmaneuver the Grim Reaper. Dozens of times each day, gravely injured patients roll into Shock Trauma's ten bays in need of urgent action to keep them alive, with little time to spare and no margin for error. Some days, Scalea might perform twelve emergency surgeries, from open-heart procedures to organ repair to bullet removal, weathering stretches of up to twenty hours without food, sleep, or breaks. Ever since he took over as head of Shock Trauma in 1997, Scalea has worked somewhere around one hundred hours per week. He manages a staff of more than four hundred people. And in each case he oversees, Scalea knows the patient's life depends on his actions. "If I do something wrong and somebody dies, it's hard to blame God," he told me.

Anyone who has watched television shows like *ER* is well acquainted with the typical scene during one of these pressure-packed trauma surgeries: doctors and nurses dash around the operating room in a frenzy, clutching items of blood-soaked medical equipment, furiously yelling "Clear!" and "Stat!" until the final cathartic moment when the resolute lead surgeon grips the fading patient by the shoulders and shouts, *"LIVE, damn you!"* Of course, these dramatizations just make Scalea chuckle. Real trauma medicine, he says, is efficient, calm, and free of theatrics. "Nobody in this hospital has ever heard me yell," Scalea said. "Ever. Because if I yell, then I've lost control of the situation. Yesterday, for example, we were operating on this terrible case and something bad happened, and suddenly the patient was pouring blood everywhere. So you know what I did? I grabbed a piece of gauze to stop the bleeding, and when everybody looked at me, I said, 'Gee, that's too bad. Here's how we're going to fix this.' And everyone was at ease. But if you scream, everyone's on edge."

Peeking in on a typical Shock Trauma case reveals a scene eerily stripped of tension; as Scalea and his team operate, you can almost imagine they're playing a board game, not snatching a patient from the jaws of death. Take the case of Larry Kiser, a carpenter in his late thirties whose brush with mortality at Shock Trauma was captured on video by an MSNBC crew. On the job one morning, Kiser fired his nail gun at a board but became puzzled when nothing emerged from the other end. As he was looking around for the missing nail, he felt a sharp sting in his chest: Kiser had shot the gun backwards, sinking the two-inch nail cleanly through his breastbone. Paramedics wheeled the ashen Kiser into Shock Trauma, where Scalea, clad in powder-pink scrubs and a New York Yankees bandana, surveyed the damage with a handheld ultrasound device. In the same tone of voice that one might use to inform someone that he'd dropped a dime on the sidewalk, Scalea told Kiser, "You've got a hole in your heart, young man." Although Kiser was talking lucidly, blood was pooling fast in his chest; Scalea believed he'd be dead in another twenty minutes unless they intervened. Yet for Scalea, it seemed like just another day at the office. Over the next two hours, the Shock Trauma squad opened Kiser's chest, removed the nail, repaired his heart, and patched him back together again, all of it happening in an atmosphere of relaxed, quiet focus, with the occasional light chuckle at some witticism.

Perhaps even more incredible than the Shock Trauma staff's nonchalance in a life-threatening medical crisis, however, is the unit's extraordinary effectiveness under routinely high levels of pressure. A couple of days after his emergency surgery, Kiser was already up and chatting pleasantly with hospital staff about his nail-gun mishap ("I asked 'em for my nail back, just as a keepsake," he reported), and in this respect he's in the vast majority: out of the nearly eight thousand mortally wounded souls who enter Shock Trauma on stretchers each year, an astounding 97 percent of them survive. A decade ago, when Scalea took over, that same figure was 92 percent. Shock Trauma's staff has become so good at working under life-or-death pressure that even as Baltimore's notoriously high violent crime rate keeps rising, they've managed to *cut* the city's murder rate.

Scalea himself presents us with something of a mystery. Every day, he works under an amount of chaos and pressure that would send most of us scrambling for a bottle of Valium, and yet he's not just unaffected by the stress—he has trouble *understanding* how anyone could see it as stressful. "When I was at Kings County [Hospital Center] in Brooklyn, I'd hear people talking about how things were so stressful, and I always said, 'What do you mean, "stressful"? That's crap,'" he told me. "One day, another doctor had to take me aside and tell me, 'You need to remember that stress is a perception. Just because you're not stressed, that doesn't mean the same thing doesn't provoke stress in the residents.'" According to the codes of human psychology, Scalea's endless work hours, massive responsibilities, and constant exposure to turmoil should play havoc on his nerves. Indeed, many residents who try trauma medicine learn rather abruptly that they're not cut out for it, which makes specialists like Scalea a rare breed; the United States produces fewer than fifty new trauma doctors each year. But for Scalea, a man whose boundless self-confidence could fill a stadium (saving a few thousand lives will do that for you), his hectic calling provides nothing but exhilaration. "I always say, 'Trauma is what you do when instant gratification takes too long,'" he said. "When someone comes in dead and you make them alive again in three or four hours, that's definitive satisfaction."

As psychophysiological phenomena go, stress is a strange customer. Generally speaking, stress is a lot like fear and anxiety—it's the body's response to a demanding, potentially threatening state of affairs—yet it's also quite different. Fear is primarily physiological and anxiety is mainly cognitive, but stress can come from the body, the mind, or both at once. Stress is the body's response to circumstances we perceive as trying, strenuous, or just plain irritating, which means that psychological stressors (like discord at home), physical stressors (like annoying sounds), and mixtures of the two (say, your screaming, tantrum-throwing three-year-old) all fuel stress. And unlike fear and anxiety, stress tends to stack up over time. Even minor nuisances can start it mounting; in one Japanese study, researchers who subjected volunteers to the sound of a cell phone ringing once a minute for half an hour found that the incessant noise shot their levels of cortisol, the hormone the body secretes when stressed, straight upward. (Number of readers who are surprised by this finding: zero.) Stress is also very much a matter of perception, as we've seen with Scalea. A single twenty-hour work shift might scar a dainty writer like myself for life, but Scalea thrives on such days. Thinking of an activity like skydiving leaves many clammy with dread, yet some need the prospect of something so thrilling just to get out of bed in the morning. Whether a potential stressor becomes debilitating or invigorating is largely a product of how we look at it.

Stress is so tightly woven into our lives today that it's difficult to believe the concept itself is barely seventy years old. In the 1930s, the Hungarian endocrinologist Hans Selye first coined the term *stress* to describe the physiological changes he noticed in rodents after injecting them with various unpleasant substances — effects ranging from muscular tension to gastric problems. For Selye, stress wasn't a uniformly bad thing; he actually used two subterms for stress, with *eustress* being the kind of stress that energizes and motivates you, and *distress* being the kind that makes you want to throw a stapler through your computer monitor. Still, research into stress has focused almost exclusively on the less pleasant side of the equation. Allow me to summarize seventy years of stress research in two words: stress sucks. As the stress expert and Stanford University neurobiologist Robert Sapolsky puts it in his book *Why Zebras Don't Get Ulcers*, "Stress can wreak havoc with your metabolism, raise your blood pressure, burst your white blood cells, make you flatulent, ruin your sex life, and if that's not enough, possibly damage your brain." And troublingly, our modern urban world is transforming into a veritable snow globe of hidden stressors. In 1999, one psychologist found that people living in New York City not only have a heart attack rate that is 55 percent above the national average — even *visitors* to the city have a 34 percent higher heart attack rate than normal.

In taxing times, most of us find our brains clogged, our bodies tense, and our nerves fraying fast. Stress and chaos make it difficult for us to function, plain and simple. But there are some among us who thrive on turmoil, who deflect stress like Gore-Tex repels water, who can plow through ten consecutive trauma surgeries and be eager for number eleven. Individuals, in other words, who have figured out how to sail through the tempestuous seas of stress and pressure without smashing into the rocks. What do these people know that we don't?

Bombs and Breakdowns

In the summer of 1940, one year after the United Kingdom officially entered World War II, German commanders hatched a plot to break the will of the British people. The plan was simple: every night for months on end, hundreds of Nazi planes would shower their bombs over London (and later other cities) until the English citizenry, in a state of unanimous nervous breakdown, pleaded with its leadership to raise the white flag of surrender. London itself was completely unprepared for aerial assault, which meant Hitler's bombers could soar over Britain at will. National psychological collapse seemed only a matter of time.

Before the air raids — forever after known as the Blitz — began that summer, British leaders steeled themselves for pandemonium on an epic scale. Expecting a barrage of psychological meltdowns, officials quickly trained fleets of volunteer mental health workers to manage the coming crisis. They established impromptu psychiatric clinics and formulated plans for evacuating the worst head cases. And as the first swarm of 364 bombers lit up the city on September 7, Hitler's psychological campaign appeared to be working perfectly; Londoners, quite predictably, were terrified at seeing their city suddenly aflame, and the overwhelming Nazi force lost virtually no planes to British antiaircraft fire. Through the winter, the Nazis subjected Londoners to almost nightly bombing, hitting them on as many as fifty-seven consecutive evenings, until, finally, the stress drove the citizenry to...

Well, to be honest, it drove them to boredom. A funny thing happened on the way to national nervous collapse: the stress pandemic never materialized. The Nazi bombing campaign was physically ruinous to the city of London, but not to the psyches of its citizens. As the British psychologist Philip Vernon later reported after collecting data on the Blitz from fifty of his colleagues, Londoners showed plenty of fear of the air raids at first. "Before the end of 1940, however," an awestruck Vernon wrote, "Londoners were generally taking no notice of sirens at all unless accompanied by the noise of planes and gunfire or bombs." Yawning citizens liked to lounge around gossiping with neighbors amid the clamor of air-raid sirens until the very last minute; they were so unperturbed about the bombings that suddenly the same officials who had worried about widespread psychological breakdown were complaining that people weren't fearful *enough*. The mental health workers found they had nothing to do. Mollie Panter-Downes, a *New Yorker* correspondent, wrote that even as the raids got worse, commuters "placidly bragged to fellow passengers on the morning trains about the size of bomb craters in their neighborhoods, as in a more peaceful summer they would have bragged about their roses and squash." Despite all reasonable expectations, Hitler's strategy to stress out the British public failed dismally. But why?

Hold that thought for a moment.

Now let's look at another example of the human response to dangerous projectiles falling from the sky. If ordinary British civilians responded to the nightly Blitz attacks with a shrug, then surely Her Majesty's most valiant soldiers wouldn't mind a little artillery fire, right? In his World War II memoir *Goodbye, Darkness*, the historian William Manchester tells of just such a fighter with whom he served, a "strutting, bullying, powerfully built sergeant major" who laughed in the face of enemy rifle fire and single-handedly destroyed Japanese machine-gun nests. Yet, as Manchester reported with a twinge of satisfaction, enemy artillery fire reduced this heroic man to a gibbering wreck. The sergeant major "just couldn't stand the strain of concentrated enemy shellfire," Manchester recalled. "Artillery turned his bowels to water." In fact, soldiers from all sides routinely reported that artillery fire was intolerably terrifying and stressful. As the World War II veteran E. B. Sledge once wrote, "To be under heavy shellfire was to me by far the most terrifying of combat experiences. Each time it left me feeling more forlorn and helpless, more fatalistic, and with less confidence that I could escape the dreadful law of averages that inexorably reduced our numbers." If you've ever wondered where the phrase *shell-shocked* came from, now you know.

So what's going on here? In both of these cases, we see large numbers of dangerous objects dropping from the sky toward humans who would rather not be blown up, yet the two groups have precisely *opposite* reactions to the stressor. Even more mystifying, it's the tough-guy soldiers who fall apart in response to shellfire, while the milquetoast civilians languidly finish their cup of tea amid blaring sirens. What gives?

The key to this mystery lies in two factors that explain much of our psychological stress: certainty and control. You might remember this duo from their cameo appearance last chapter, and as with anxiety, the formula here is quite basic. The more certainty and control we think we have about a potentially threatening situation, the less stress we will feel. Interestingly enough, perception is all that counts with this. You don't actually need to have perfect certainty or total control over how things will pan out; you just need to *believe* you have them. In the lab, if you give a subject the illusion that he can reduce the intensity of a stressor (like an electric shock) by pressing a button on cue, he shows far less stress than someone not given the button option—even if pressing the button does absolutely nothing. Stress shrinks, then, when we perceive certainty and controllability in a rough spot, and it mounts when those factors are absent.*

Now let's look a bit more closely at those two examples above. When the Nazis sketched out their psyops tactics for the Blitz, they committed a terrible blunder: in conducting their bombing raids at the same time every night, they gave Londoners a total sense of predictability. No one had to worry if the planes would come that night; they simply knew they would, just like Germany's famously punctual trains. Thus, the steady diet of the same bombing run over and over simply made the Londoners habituate to the siege. (In the British countryside, where the bombings were sporadic and unpredictable, the public grew far more stressed.) At the same time, having the option to flee to underground shelters gave Londoners a sense of control over their odds of survival.

Our shell-shocked soldiers, on the other hand, had the luxury of neither certainty nor control. Mortar fire was utterly random and arrived on an unpredictable schedule. Troops on the front lines had no fortified shelters to escape to, they couldn't fight back, and they couldn't even run away, since a shell was just as likely to hit them while they were running as while they were sitting still.

---

*The same caveat from the last chapter's eight mistakes still applies, mind you. While a sense of certainty or control alleviates anxiety and stress, seeking them when they're impossible to attain is a surefire recipe for psychological turmoil, like if you tried to gain control over the weather.

With no predictability, they always felt vulnerable. With no control, they always felt powerless. Intense stress was the result.

Because fighting in a war zone is one of the most stressful jobs imaginable, the experiences of soldiers during conflict tell us much about the landscape of human stress. Our first lesson is this: We all have a breaking point. No soldier can endure combat stress indefinitely without melting down. This news isn't too surprising to us today, but it certainly was to the mustachioed, stiff-backed military leaders of the early twentieth century. Shell shock, after all, is a distinctly modern phenomenon. In the preindustrial age, all battles had natural limits: before the advent of automatic weapons and artillery, fighting was physically demanding and it happened up close. The two sides could only take so much before packing it in for the day. This formula changed in World War I, when a slew of devastating new weapons—machine guns, artillery, tanks, mustard gas—turned war into something fought day and night in the trenches. On the front, the stress never let up, and it showed. Suddenly, soldiers were burning out and breaking down. During World War II, the U.S. Army sometimes had to discharge its "psychiatric casualties" faster than it could draft new recruits in.

It wasn't until the Second World War, however, that psychologists began to quantify the limits of human endurance under the intense stress of combat. In 1944, the American psychologist John Appel surveyed patients at shell shock centers and determined that a soldier could take only seven or eight months of combat before becoming "so overly cautious and jittery that he was ineffective and demoralizing to the newer men." Another survey during the Normandy assault found that men could stand only thirty days of continuous fighting; after sixty days, 98 percent of troops exposed to such conditions were "emotionally dead" from stress. And what about the 2 percent who didn't break down? According to psychologists' surveys, these men fit a "psychopathic" personality description before they ever entered the war.

But even as military psychologists discovered that each soldier's spirit was an exhaustible resource under fire, they also saw that some troops felt paradoxically low levels of stress in times of mortal danger—and it all traced back to certainty and control. Consider fighter pilots. According to the 1945 report "Men Under Stress," the mortality rate for dogfighters was among the highest in the military; the pilots knew that half of them would be killed in action. Yet fighter pilots also enjoyed wildly high job satisfaction, with 93 percent of them claiming to be happy with their assignments. And why should they be so content? Because fighter pilots felt they were in complete control of their fate. They could maneuver however they liked through a huge airspace, and they believed, to a man, that their piloting skill would determine their survival, not luck. (Bomber crews, who had to stick to a prescribed flight course no matter the intensity of enemy flak, felt far more stress despite having higher odds of survival.) Similarly, the 1943 report "The Moral Effect of Weapons" found that infantry troops feared enemy armaments not according to how likely they were to be wounded by them, but instead according to their predictability- and control-related perceptions about the weapons. Soldiers lived in terror of dive-bombers and mortar fire, both of which attack suddenly and offer little opportunity to fight back, but they felt more ambivalent about statistically deadlier machine-gun nests; at least they could shoot at enemy gunners. This is why rigid routines and menial tasks are such a vital part of military life: they give soldiers a sense of predictability and control. In a war zone, even something as dull as counting ammunition can make a soldier feel he's loading the odds of survival in his favor.

Few of us are likely to experience the stress of an artillery barrage anytime soon, but the lessons of the battlefield apply in life's lesser war zones as well—places like the Poulsbo, Washington, home of Courtnee and Mike Stevenson. In March 2006, Courtnee gave birth to a set of quintuplets, Aniston, Belle, Camilee, Scarlett, and Weston, who joined the couple's three-year-old daughter, Lilli. After leaving the hospital, the Stevensons repaired to (steel yourself) their nine-hundred-square-foot mobile home, five squealing newborns and toddler in tow, to begin their frenzied new life. "From the beginning, the stress was unbelievable," Courtnee Stevenson told me. "Having quintuplets isn't like the stress of a newborn times five—it's that stress to the fifth power." The Stevensons first tried to take care of their quints alone, but soon found that the physical laws of the universe prevented them from being able to handle five newborns in constant need of feeding, changing, and attention. They recruited a staff of more than forty volunteers to take shifts with the quintuplets around the clock. "Sometimes you literally couldn't walk through our house, there were so many bodies packed in," Stevenson recalled.

But over the last three years, Courtnee Stevenson has reshaped herself into Supermom, single-handedly managing the chaos and unpredictability of her toddler army. "You never know what kind of day you're going to have," said Stevenson, who, when I spoke with her at her much larger current home, had been up ministering to pairs of feverish kids for the last few nights. (Fittingly, a blog post she wrote about this began with the line, "If you saw me right now you would have thought I have been to war.") "Every day is stressful here. The kids climb, they conquer, they make tons of noise. There's no downtime. But you get used to the chaos." Stevenson is now a master of advanced mom techniques like eating dinner standing up, without looking at her plate, but she says the true turning point in better handling quintuplet stress came when she instituted the same strict routines seen in the military. "We have a very tight schedule," she explained. "The mornings are the hardest, because I have to get six kids to school by nine a.m. So I take them two at a time, dress them and feed them. And they know what's expected. They know that if they get in the car seat, they'll get a treat. It's almost like training animals." The routine reduces the unpredictability of overseeing five three-year-olds, each one a determined little cyclone of kinetic energy, and gives Stevenson a sliver of control over things. "The schedule is so structured that I always say my life is like *Groundhog Day*—the same thing every day," she said with a laugh.

No matter where we look, the same broad idea behind stress applies: a perception of uncertainty or lack of control intensifies the impact of stressors, whereas predictability and power allevi-ate that impact. Take the humble traffic jam, for example—a seemingly trivial annoyance that hides serious health conse-quences. Naturally, you already know that being stuck in traffic is stressful. In surveys, people routinely rank it as their top daily stressor; even seeing the word *traffic* just now may have forced you to involuntarily fling this book to the ground in blind rage. But as a team of German medical researchers found in a 2009 study, this stress is nothing to scoff at. Being stuck in traffic, they learned, more than *triples* your chances of suffering a heart attack over the following hour. Why, we might ask, does this cause such tension? Aren't we simply *sitting there* in the car? The stressor in traffic clearly isn't the sitting, though; it's the sense of powerless-ness and uncertainty that traffic jams induce. Encased within an unmoving armada of cars, we have no idea when the gridlock will let up, and we can do nothing to change things—no predictabil-ity, no control, high stress. In fact, small uncertainty-reducing travel adjustments such as booking a direct flight or taking an express train to work instead of a local one can do wonders for your sense of well-being.

At the office, the rules of stress are no different. In study after study, researchers have found that the most stressful occupations are those in which employees must deal not just with high demands, but with little control over their workdays. As the Rockefeller University stress researcher Bruce McEwen explained to the *New York Times*, "The people who are under someone's thumb, who are low-ranking and don't have any decision-making [power], these are the people who always experience more anxiety." Studies of classical musicians, for example, have long shown higher job satisfaction in small chamber groups than in large symphonies, because orchestra members have so little autonomy; conductors have historically held so much control that musicians haven't even been able to take bathroom breaks during rehearsal without permission. (Years back, some psychologists proposed an "executive stress syndrome" to honor the extreme strain that bosses endure, but it turns out that executives display *fewer* stress symptoms than their underlings, because they have both greater control over their work lives and the ability to displace stress onto subordinates.) In the chronically tense modern American work world, some benevolent employers have begun offering company-subsidized stress-reduction perks like yoga and massage to keep their workers from keeling over from strain.* But as Lawrence Murphy, a psychologist at the National Institute of Occupational Safety and Health, has concluded, these practices are merely a Band-Aid for work stress. They give short-lived relief, but they don't address the serious underlying issues causing the tension: lack of control, unpredictability, high demands, and bad management.

---

*In Japan, enough people have actually died from work stress that there's even a term for it: *karoshi*. Officials there have had to set up a national *karoshi* hotline, publish *karoshi* self-help books, and establish a legal precedent for compensating spouses of *karoshi* victims.

To reduce stress in an enduring way, psychologists say, we need to shift ourselves from a sense of jumbled helplessness to one of self-assurance. They call this confident attitude a strong "internal locus of control"—the sense of being in the driver's seat of your life, as opposed to feeling buffeted about by forces beyond your influence—and it correlates positively with many measures of mental well-being. Building up an internal locus of control is simpler than it sounds: we can fortify it with easy measures like brainstorming plans of action when we feel stuck (to show that we do have options and agency after all) or reminding ourselves that *we* created our major and minor achievements in life, not fate or luck. Again, belief is all that counts; you needn't actually have power over everything, just a stalwart sense that you can act to shift things in your favor. So ultimately, it really is true that stress largely relies on perception. Which helps explain why those with cool heads can wander so unworriedly into a maelstrom of uncertainty: in chaos, they see a whole different world than the rest of us see.

Perception Is Reality

In my imagination, at least, an air traffic controller's first day on the job might look something like this. A slightly balding middle manager, mug of coffee in hand, escorts the new initiate to his workstation atop a control tower, its panoramic windows overlooking a vast plain of airport asphalt. Pulling out a chair at an empty workstation, he begins his friendly initiation speech: "So first of all, welcome. Here's your radar scope and headset. You know the drill: keep the planes separated, blah, blah, blah. Before you begin, though, just keep one thing in mind. If you commit a little blunder over here, then—oh, how can I phrase this?—hundreds of people may die in a horrific flaming wreck. Just a heads-up. Good luck!"

I'm exaggerating, of course, but given the gravity of what air traffic controllers manage each workday, it's tough not to see the job in such weighty terms. On any given air traffic control shift, tens of thousands of passengers unknowingly depend on a controller's vigilance to keep them safe in a sky increasingly clogged with aircraft. "There's just a constant stream of decisions and consequences that go on throughout a controller's workday," said Ned Reese, who was the Federal Aviation Administration's head of training for air traffic controllers for a dozen years. "The clock never stops. These are aircraft moving at 450 knots [about 515 miles per hour], and it's not one of them—it's thirty-five of 'em at once, headed in all different directions." For decades, stories have proliferated about high burnout rates within the profession, and the strain only looks to get worse. According to figures from the Bureau of Transportation Statistics, the average number of domestic passenger flights each day has risen by a third since 1996, but the number of working air traffic controllers has *declined*. With dozens of juggling balls in the air at once, each one signifying hundreds of lives, it's a hell of a stressful profession.

Or so you'd think. In reality, Reese says, the idea of air traffic controllers breaking down from strain has been blown out of proportion; one 2006 study found that burnout was far more common among construction workers and even journalists.* The reason for this is revealing. Often, we think of hiring for jobs according to the following formula: first, you find the person with the best skills for the job, and then you teach them to manage the stress of the profession. The stress is taken as a given, an inevitability. But when Ned Reese took over as director of air traffic training at the FAA in 1992, he and his colleagues decided to look at the question from

---

*Number of journalists who are surprised by this finding: zero.

a different angle. Instead of instructing trainees in how to fend off job stress, what if they looked for people who didn't *experience* air traffic control as stressful?

As Reese likes to point out, almost anyone can learn and apply the basic rules of air traffic control, but that ability isn't the whole story. "If you give me a bright person and enough time, I can teach them all the regulations and protocols they need to know," he told me. "But the real test is when the clock suddenly starts: now you've got to take those rules and perform without having to stop and think." For decades, Reese says, the FAA selected air traffic controllers through tests that measured cognitive aptitudes like math ability and memory skills, both of which are necessary for air traffic control work but not sufficient; they demonstrate that you *can* do the job, but they don't show *how* you'll do it. So when Reese and his colleagues set out to formulate a new air traffic test in 1996, they decided to look for a broader picture—not just aptitudes, but intuitive cognitive skills as well. How mentally flexible was the trainee? How did she react to a changing environment? How easily did she move information from short-term memory to long-term memory? These weren't rules one could memorize; they were built-in cognitive skills, and they affected not just how well a trainee could perform but how she perceived the world.

The test they ended up with, the Air Traffic Selection and Training battery, gave Reese and his colleagues a fuller idea of how a trainee would approach the job. The difference between a candidate with the right formula of innate skills and one without it, he says, is readily apparent. "Let's say you put two controllers side by side, and they're working the same level of traffic," he explained. "One would be sitting down, then standing up, then stretching out, writing and rewriting things, clearing and reclearing aircraft, pointing at the scope—just a lot of physical activity. And that would be safe. The controller next to him would be working the same traffic, but you wouldn't know he was even awake from watching him from behind. He might make only as many transmissions as absolutely necessary, and he's at the same level of safety. So who's going to have a longer, healthier career?" When those two controllers look into a radar scope glutted with speeding aircraft, they see two different scenes. The first sees a chaotic, hazardous situation that he must stay on top of through constant activity, which translates to stress. But the second controller sees the ever-changing map as something else. "It's like a puzzle to them," Reese said. "They look forward to it, and they do the job without even thinking. They walk out of the control room after eight hours and they're tired but not wasted." The difference comes from how they look at change and ambiguity. One man's stress is another man's challenge.

Not long after I spoke with Reese, I called a man who, I felt confident in assuming, couldn't *not* experience crippling job stress, given his line of work. For thirteen years, Frederick Lanceley worked as the FBI's senior crisis negotiator, which meant that whenever a deranged person decided to put a gun to someone's head and start making demands, he was on the short list of people to call. Lanceley's bio attests that he has negotiated "several hundred hostage, terrorist, barricade, suicide, aircraft hijacking and kidnapping cases worldwide," each of them a study in tension and fear, with only a delicate, paper-thin barrier of sanity holding off catastrophe. The hostage negotiator must walk a tightrope of terror and stress, ride a roller coaster of emotion, soar on a gale of pure...wait—*what?* "It's actually kind of boring," Lanceley corrected me from his office near Atlanta. "You're there for hours and hours, and there's nothing happening. I've been asked many times about making a movie about my experiences in crisis negotiation, and I always ask, 'What are you going to put in the movie?' People are interested in tactical teams going in and sniper shots, but in a good negotiation, nothing happens."

Hostage situations, Lanceley explains, tend to follow the same trajectory. At first, everyone on the scene is tense: someone's life is in danger, and no one knows what's going to happen. Usually, the hostage taker is highly emotional, which only seems to make the situation more precarious. But within an hour or two, Lanceley says, things become remarkably dull. "If you're in emotional crisis, you can only keep that up for so long," he said. "You can be raging, screaming angry, but you have to come down eventually." Lanceley compares his role in the negotiation process to that of a therapist. "For the first couple of hours, the hostage taker isn't ready to listen to anybody, so what you do is keep him talking," he said. "You let him blow off steam, indicate that you're listening and that you care, and tell him that nobody's going to hurt him. After four or five hours, the emotions or the drugs wear off and it ends." When demands come out, Lanceley takes them as a good sign; it means the hostage taker isn't really there to hurt anybody.

Amid the sitting and waiting of a crisis, Lanceley seldom felt the stress one would expect from his job. "I've talked to plenty of cops who have told me, 'Negotiating with the bad guy was easy — it's dealing with my boss that's hard,'" he explained. When he *did* feel the stress mounting, Lanceley employed a simple cognitive trick. "It might sound less than heroic, but one of the ways I keep my cool under pressure is by reminding myself that I'm not in danger," he told me. "I'm not going to die here. Somebody else might, but not me. It's not heroic, but it's practical."

None of this is to say that air traffic control or hostage negotiation or trauma surgery isn't stressful; for most people, each clearly would be. The point here is that the stressful aspects of these jobs aren't a given. Often, our stress level depends more on how we perceive a set of cognitive variables than on the specific demands of that situation. When Ned Reese's ideal air traffic controller peers into his radar scope and sees three dozen planes hurtling through his airspace, he interprets this constant state of change not as threatening but as challenging. When Frederick Lanceley begins negotiating with an emotionally charged hostage taker, he experiences this state of affairs not as volatile and uncertain but as having a predictable arc. When Thomas Scalea, the trauma surgeon, strides up to a gurney that holds a man with a nail puncturing his heart, he sees the case not as unmanageable or dire, but as a predicament over which he can exercise quite a bit of control. Their cool-headedness hinges on what they see when they peer into the chaos, things like challenge and possibility. So how did they come to see things the way they do?

Sailing the Three C's

The ability to keep your mind straight through stressful circumstances is at least partially biological, and for the starkest proof of this, we turn again to the U.S. military (motto: "Stressing you out since 1775"). For the past fifty years, the toughest of the tough in the U.S. armed forces have trained at the John F. Kennedy Special Warfare Center and School at Fort Bragg, North Carolina—a.k.a. "the Schoolhouse"—to join the ranks of the Special Forces, which includes legendary outfits like the Navy SEALs and Green Berets. It's an elite group, comprising just 10,000 soldiers, and to get in, recruits must train under grueling conditions for two to three years. The Schoolhouse regimen revolves around testing stress resilience: recruits face pain, exhaustion, sleep deprivation, and, crucially, *uncertainty*. As one trainer told the *New York Times*, "We never inform them what they're going to do, how long it's going to go on. We set the conditions for maximum ambiguity from the start." If recruits can't show flexibility in the midst of confusing, ever-changing situations, they get weeded out. And as the finishing touch to the Schoolhouse experience, Special Forces trainees endure an infamous three-week exercise known as SERE, for "Survival, Evasion, Resistance, and Escape." The military calls it a resistance-training laboratory, which is essentially a euphemism for "torture center." For those three long weeks, trainees endure the psychological and physical abuse they might receive as prisoners of war, complete with interrogations, forced isolation, malicious guards, even "advanced techniques" like waterboarding.

For more than a decade, the Yale psychiatrist Andy Morgan has researched Special Forces trainees at Fort Bragg to try to discern why some recruits function so much better than others under stress. "I do the Jane Goodall thing, just following them around, collecting spit and blood," Morgan told me. "They seem to like spitting. I don't know why. It's very primitive." One of the first things that Morgan and his colleagues' research revealed was that SERE training was one of the most brutally stressful experiences ever studied by scientists; as he and a collaborator wrote in one 2000 report, "Recorded changes in cortisol levels [the major stress hormone] were some of the greatest ever documented in humans." The training was even more punishing than undergoing open-heart surgery. Yet Morgan also found that many Special Forces recruits in the SERE program preserved an amazing amount of mental clarity under stress; quick cognitive tests showed that they could keep it together despite the physical toll. "They weren't necessarily happier about the stress," Morgan explained. "They said, 'Yeah, it sucks—but I'm not rattled.'" When Morgan compared these poised soldiers' blood tests with those of more scattered and distressed recruits, he noticed that the clear-minded trainees were producing significantly higher levels of "a goofy little peptide called neuropeptide Y"—often as much as a third more of it. It was like a layer of bulletproof Kevlar protecting their minds.

"Neuropeptide Y seems to specifically counter the negative

impacts of high arousal in the brain, like becoming jarred and scattered," Morgan said. "It helps people stay actively engaged in their environment, even if their levels of adrenaline and cortisol might in fact be higher than the people who aren't doing as well." The effect of neuropeptide Y is so pronounced, Morgan says, that he can tell if a soldier is Special Forces or not simply by taking a look at a blood test.* Although Morgan's "goofy" neuropeptide plays a significant role in our ability to function under stress, Morgan claims that the way we talk to ourselves about stressors is equally important. "How you frame something in your head has a great deal to do with your neurobiological response to it," he told me. "Once you start saying to yourself, *Oh my God, this is awful,* you begin releasing more cortisol and start this cascade of alarm. But when you say to yourself, *I know what to do here,* or see things as a challenge, then that turns into a much more positive response."

If there's one group of people who epitomize this upbeat view of perilous, highly demanding situations, it's that clan of unique individuals—by which I mean "lunatics"—who spend their leisure hours scaling sheer cliffs, hurtling down icy slopes, and leaping off things of various heights: risk-takers. To those of us who approach life with a strong preference for keeping our limbs firmly attached to our bodies, dedicated thrill-seekers truly do seem unhinged. For the Temple University psychologist Frank Farley, however, risk-takers aren't mysterious head cases at all— they just see the world in a distinctive way. Farley has spent his

---

*Another example of biological stress resistance is the so-called resilience gene, 5-HTT, which comes in two forms: short and long. We each carry two copies of the 5-HTT gene—one from each parent—and research has shown that those with two short 5-HTT genes are two and a half times more likely to get depressed after stressful life events than those with two long 5-HTT genes. Most of us, though, have one of each.

entire forty-year academic career researching the differences between those who thrive on risky, demanding activities (whom he calls the Big T personality type, with the *t* standing for *thrill*) and those who veer toward the safer, more predictable end of the spectrum (whom he calls the small t type). "The more I've looked at risk taking, the more it seemed like a key ingredient in life," Farley told me. "There's a wonderful quote from Helen Keller, who said, 'Life is either a daring adventure or nothing.' I've been very impressed with that kind of thinking all my life."

The question of *why* we might view life's risks so cheerfully, Farley says, boils down to how we interpret things like uncertainty, novelty, and change—factors that show up in all stressful activity. "Small t's are very enamored of security, stability, predictability," Farley explained. "When I ask small t's, 'What's your number one goal in life?,' a very frequent answer is 'To lead a secure life.' But Big T's are the opposite. When I ask them the same question, they'll often say, 'To lead an exciting life.' They would love an ever-changing world. The small t's would love a never-changing world." Whereas uncertainty and flux often bring stress to those with small t personalities, Farley's research shows that risk-takers associate chaos with "excitement and opportunity." They also tend to believe that they can handle whatever comes up, that their destiny lies in their own hands—the strong internal locus of control we discussed earlier. And they like to test their boundaries, even if it means occasionally crashing and burning. These differences add up to a unique view of life's risks, Farley explains. "It's not as if they have no fear," he said. "They know there's risk, but it's meaningful to their life. Big T's don't want to die in bed with tubes running into their body. They want it to be while doing the grand thing."

No one is saying that being one of Farley's small t's is *bad*; in fact, I consider myself to be something of a legend within the thrill-averse kingdom. The risk-taking personality type is a continuum, and most of us fit somewhere between the two extremes. But with a modicum of work and a shift in how we frame life's stressors, anyone can become more flexible in the face of change and uncertainty. Just like with anxiety, our genes and inborn dispositions make up only part of the story of how we cope with stressful circumstances. A little effort to build resilient new cognitive and behavioral habits can go a long way.

At the forefront of the stress resilience training movement stands the UC Irvine psychologist Salvatore Maddi, who has been trumpeting his findings on building hardiness for a quarter century. Maddi fell into the stress resilience field almost by accident. Back in the mid-1970s, when he was teaching at the University of Chicago, Maddi began a long-term psychological study of 450 management employees at Illinois Bell Telephone—but he wasn't even studying stress at first. "I was actually consulting for them on how to help their staff become more creative," he recalled. The research entailed yearly psychological and medical evaluations of the Illinois Bell managers, but six years into Maddi's study, an unexpected corporate cataclysm jolted his data: the federal government broke up the Ma Bell telephone monopoly. Suddenly, tens of thousands of Bell employees found themselves trapped in one of the most severe corporate shake-ups in history. Jobs changed abruptly or disappeared completely; rules and protocols shifted constantly; and, most important, no one really knew where, how, or *if* they'd be working for a Bell company in a month's time. It was, in short, a monumentally stressful time for a lot of people.

As the Ma Bell maelstrom raged on, Maddi and his colleagues kept gathering data. "What we found was that six years after the upheaval, two-thirds of the people in our sample fell apart," Maddi told me. "There were heart attacks, there were strokes, there was violence in the workplace, there were suicides, there were divorces—everything you can imagine." Predictably, the stress and uncertainty did these people no favors. But what about the other third of the managers? How did they fare? "The other third of our sample not only survived—they actually thrived," Maddi said. "Their capabilities came forward even more than before. They had more energy, felt a greater sense of fulfillment in their lives, had fewer illness symptoms, and actually performed *better* than before." Cataclysmic work stress somehow brought out the best in these resilient workers, regardless of whether they stayed with the company or lost their job.

Because Maddi and his fellow researchers had twelve years' worth of exhaustive psychological data from before and after the corporate shake-up, they could now ask the all-important question: What was the difference between those who crumbled and those who flourished? Maddi found three common themes, a trio of attitudes about stress that he has dubbed the "Three C's." First, the hardy Illinois Bell employees all displayed *commitment*; instead of running away as stress mounted, they remained involved with the world around them, actively pursuing goals no matter how tough things got. Second, they showed a strong sense of *control*; they recognized that even in the roughest circumstances, they were never helpless. And finally, these resilient workers saw *challenge* in their ever-changing world; they could mentally transform a crisis from a threat into an opportunity for growth. Of course, these three factors echo plenty of what we've seen already about dealing well with stress—the internal locus of control in risk-takers, the welcoming attitude toward change and uncertainty in air traffic controllers—but Maddi has grouped his Three C's of commitment, control, and challenge under a single umbrella term: *hardiness*. "People high in hardiness realize that life is by its nature a stressful phenomenon," he explained. "What you have to do is use stress and change to grow and develop."

As we all know, this is no easy task, yet Maddi is adamant that any one of us can instill these resilient attitudes. "We know we can train people to increase their hardiness," Maddi said. "It involves showing people how to deal with their stressful circumstances through problem solving, coping skills, social support, and effective self-care—rather than just giving up." Hardiness trainees learn to identify their stressors and to challenge their negative beliefs about them, like the idea that they're helpless to change a tough situation or that it's safer to give up on a problem when it grows difficult. They practice setting goals and taking small steps toward accomplishing them, slowly building up their sense of control over their own destiny. In talk sessions, therapists help them to put their stressors into perspective: they see how turmoil often leads to accomplishments and personal growth and how change is an essential element of life. They build a habit of seeking support from others when the going gets tough, and of embracing problems as new challenges. The training has so far proven effective in helping people navigate through life's stresses; Maddi likes to cite one study in which university students who took his hardiness training course ended up graduating with higher grade-point averages than those who didn't. The U.S. Army is so intrigued by hardiness training that it announced in 2009 that all 1.1 million of its soldiers will be required to complete a stress resilience course.

So by developing a stronger commitment to pursuing our goals, fortifying our internal locus of control, and learning to view dilemmas as opportunities to grow, we can build resilience to stress. But that said, it's also clear that when a patient is bleeding to death or a house is being consumed by fire right in front of your eyes, no amount of optimism about change is going to magically present you with the right *decisions*. To perform at a truly expert level under stress, you need more than the right attitude; you need to do your homework first.

The Rest of the Iceberg

In 1984, the psychology researcher Gary Klein won a grant from the U.S. Army to study decision-making under stress in firefighters. In his grant proposal, Klein offered pages upon pages of impeccably researched theories about how firefighters made tough decisions in pressure-packed situations. All of the evidence, he wrote, demonstrated that expert firefighters must mentally generate a set of options for how to handle a given predicament, compare them side by side, and then proceed with the best one. When he got to Cleveland, where he'd be studying a fire department up close, Klein would just have to gather the data to support his hypothesis before marching home in triumph.

Then Klein actually talked to the firefighters, and what they told him made his research career flash before his eyes. In one of the first interviews he conducted, Klein asked a seasoned fireground commander to describe a few examples of how he made difficult decisions under stress in the past, and the man replied with a jaw-dropper: "I don't make decisions. I don't remember when I've ever made a decision." In a time of crisis, the commander explained to a stricken-looking Klein, you have no time to weigh options—you just look at the fire, see what to do, and act. The revelation decimated Klein. "It was a terrible moment, because we were trying to examine how they compared options to make decisions, and they told us they weren't doing that at all," Klein recalled. "And it wasn't just one person telling us that—we heard it again and again. Emotionally, it was devastating, but that devastation was important because it allowed us to break free of the fixation that this was the only way to make decisions under pressure."

Klein and his colleagues had a puzzle on their hands: if the veteran firefighters said they didn't scheme or deliberate in order to arrive at the right course of action, then how could they possibly know what to do? To find a solution, he pored over dozens of incident accounts, searching for common themes in how the firefighters talked about their actions under stress. Soon enough, an insight emerged: the elite firefighters had seen so many fires over the years that a web of patterns had been etched into their subconscious minds, lending them incredibly solid instincts. Just as none of us has to consciously mull over how to put on a pair of pants (unless they're particularly complicated pants), these firefighters intuitively knew what to do in dicey spots. "There's no magic here; it's just experience," Klein explained. "What ten or fifteen years of experience buys you is that you build up a large repertoire of patterns, and even though you're not likely to see a situation that *exactly* matches one of your patterns, it'll be close enough most of the time. That's how experience expresses itself." When a fire commander sees a building with a billboard affixed to the side go up in flames, for example, experience immediately reminds him of the times he's seen billboards suddenly plummet to the ground below, and he gives orders to clear out the area without having to think. Experts test out these decisions by running a quick simulation in their heads of how it'll play out—not deliberative thinking, but a sort of intuitive visualization. If it works in their mind, they'll do it. Experience paves the way for instinct.

Klein's model combining experience-based intuition with mental simulation has proven true in studies of everyone from fighter pilots to chess grand masters to trauma surgeons like

Thomas Scalea. After all, Scalea might be blessed with a personality type that sees nothing stressful in having the patient in front of him "circle the drain" (as they so artfully put it at Shock Trauma), but to do his job well, Scalea still needs to be able to make sound medical decisions with little information and zero time to spare. "Some people need certainty, or they need time to think," he told me. "They say, 'Now that I've been presented with these symptoms, I need to go sit in the corner and consider my possibilities.' That's great if you want to be a different kind of doctor—but not this kind." Scalea credits a few factors for his ability to make reliably good decisions under life-or-death pressure. First, there's the simple matter of concentration; nobody wants their surgeon to be contemplating her tennis backhand while she's simultaneously patching up a cherished internal organ. "In order to perform well under stress, you need to be able to concentrate only on the task at hand," Scalea explained. "You need to forget that if you make a mistake then this person is going to die, or you'll become paralyzed." Second, there's his tremendous faith in himself. "You have to believe you're good enough to beat the odds, every time," he told me. "When someone comes in looking really bad, you have to say, 'I don't give a shit if this is a one in a thousand shot—*this* is that one in a thousand.' Otherwise, why even start?"

And finally, just as Klein would predict, Scalea says his decision-making under pressure keeps improving as experience sharpens his intuition. "There's a lot of pattern recognition here, and that's something you can't teach," Scalea said. "I've done this for so long now that everything's almost on autopilot." Recently, a man had arrived at Shock Trauma with a severe gunshot wound to his chest, and conventional wisdom dictated that Scalea should patch that area up first. But he noticed that the blood gushing from the wound was suspiciously dark. "I just *knew* it was his liver," Scalea recalled. "I said, 'You know what? The bullet somehow went through his liver, and this blood is going up through his diaphragm and out his chest.'" Because Scalea would have to ignore the man's grave chest wound and make an incision in the healthy-looking abdomen, the patient could have died if his instincts were wrong. Yet, as is usual with so many Shock Trauma cases, they weren't; Scalea's intuition saved the man's life. Later, Scalea's baffled team asked him how he knew about the liver wound. "I don't know," Scalea said with a shrug. "I just knew."

For the psychologist Anders Ericsson, Scalea provides a perfect case study in how elite performance under stress really comes about. Ericsson, a kindly, bearded Swede who teaches at Florida State University, is one of the world's leading experts on *experts*—specifically, what separates those who perform at a superior level from those who don't. At Florida State's Human Performance Laboratory, his team of researchers creates elaborate simulations to test subjects' poise in taxing situations. They send nurses into fully outfitted hospital rooms to see if they can keep a mysteriously flatlining robot patient alive. They park police officers in front of huge video screens, outfit them with eye-movement recorders and laser-tracked handguns, and then see how they deal with crisis simulations—when they unholster their guns, where they look, and so on. One consistent conclusion from this research is something that psychologists have long suspected: people who perform well under stress glean more information from their environment than those who perform poorly. Ericsson's Florida State colleague Paul Ward, for instance, has found that elite soccer players make better on-field decisions than lower-level players in large part because they're paying more detailed attention to what's going on around them. (Under stress, our natural tendency is to forget about our surrounding environment and focus instead on the immediate threat. The more people can resist this and pay attention to what's really happening, the better they do under fire.)

But Ericsson's signature finding over his long research career has been on the value of practice in functioning at an expert level, and his work adds an important caveat to Klein's research. Not all experience will help improve decision-making under stress; you need a certain *kind* of experience. "I've been unable to find any evidence showing that experience has any benefits unless people pay attention to feedback and actively adjust," Ericsson told me. "You need to challenge yourself and refine what you're doing."

For decades, psychologists have preached a "10,000-hour rule" that applies to the loftiest levels of expertise: to reach an elite level of skill in any pursuit, the rule says, one must first put in 10,000 hours of practice—around three hours a day for ten years. (This finding most recently garnered attention after Malcolm Gladwell featured it in his 2008 bestseller *Outliers*.) Michael Jordan and the Beatles, they say, practiced their craft for 10,000 hours before becoming megastars. Yet Ericsson often sees highly experienced nurses and police officers in his lab crumble under pressure, and the culprit isn't necessarily their disposition toward stress—it's the *type* of experience they've accrued. Experience needs two characteristics to be effective: it has to be *challenging*, focusing on your weaknesses, and it has to include *feedback* that allows you to fine-tune your approach. Otherwise, Ericsson says, you're just treading water, repeating a familiar skill instead of progressing. One 2003 study of figure skaters, for example, showed that the elite skaters who excelled in pressure-packed competitions spent 68 percent of their practice time rehearsing tricky jumps; the next group down logged only 48 percent of their time doing jumps. Practice, Ericsson says, doesn't necessarily make perfect; *tough* practice makes perfect.

To Ericsson, the job description of a trauma surgeon like

Scalea supplies a perfect template for building optimal decision-making in tense times. The feedback Scalea gets, for example, is often alarmingly direct. "With most things doctors do, you don't usually get immediate feedback," Ericsson said. "If someone comes in and you fail to notice something critical, the error in your diagnosis may not come out for some time; you can't learn from it because you don't remember what you thought when you saw the patient originally. Now, with surgery, if you cut something incorrectly, you may have an artery blow right in your face." And the effect of this viscerally instant response is bolstered by the huge quantity of unique, challenging surgeries Scalea performs every day. Scalea's broad range of experience allows him, like Klein's firefighters, to arrive at astonishingly solid decisions based on hard-won intuition, and his ability to focus on the task at hand ensures that he'll see details—such as the darker blood of the gunshot victim—that others might miss. Over time, this blend produces an efficient and highly intuitive surgical machine under fire.

When we see heroic, cool-headed characters coming through under stress, delivering sound decisions and performing smoothly, we often see their poise as primarily a product of innate character. Serena Williams was simply born to hit clutch tennis shots, we say; the cop receiving a medal for valor pulled through because of his tremendous heart. But though disposition does play a role, it's only one side of the story. "There's a lot of mystique about these people who are able to deal with stress in an uncanny way, yet what you actually see with them is just the tip of the iceberg," said David Eccles, one of Ericsson's colleagues at the Human Performance Lab. "What you're unable to see is this huge iceberg under the water that constitutes all of the experience and practice and thought that has been put into how to deal with that stress." You can be born with the most unruffled personality type in the world, one that sees limitless possibility in chaos and that thrives on uncertainty—yet only arduous, plentiful experience can give you the rock-solid instincts you need to make good decisions under fire. As Klein put it, "There's no free lunch here. There's really no substitute for experience."

Meet Barney Google

If ever there was a group of American soldiers who seemed fated to suffer lifelong post-traumatic stress, it was the band of 566 Vietnam veterans who endured imprisonment at the infamous "Hanoi Hilton" in the late 1960s and early '70s. During stays lasting as long as seven years, these American prisoners of war suffered relentless mental and physical torment from their Vietnamese captors: beatings, solitary confinement, torture, interrogation. Some prisoners were so malnourished that they lost a hundred pounds over the course of their captivity. Uncertainty and lack of control pervaded their lives; they had no autonomy, no information about the outside world, and no idea if they'd ever be released. So naturally, when the Vietnam POWs finally made it home from their ordeal in 1973, many feared for their mental health. This concern spurred the navy to keep tabs on the POWs' psychological fitness for the next twenty years, with the hope of gleaning new insights into PTSD.

In 1996, these researchers revealed the surprising result of their two-decade study: virtually *none* of the former prisoners had developed PTSD. In fact, these POWs, who had suffered so horribly, were no more likely to have PTSD than Americans in the general population—a rate of about 4 percent. (By contrast, previous studies had found PTSD rates of 50 to 90 percent among POWs.) Many of them had moved on to successful careers in business and politics, with John McCain being an obvious example. Psychologists suggested a variety of reasons for their startling resilience: like McCain, a disproportionate number of the POWs were mentally hardy, gung-ho aviators whose planes had been shot down, and they tended to be more educated, more mature, and better trained than the average grunt. The POWs themselves spoke of their religious faith and their ability to keep communicating with one another through secret tap codes—even in solitary confinement—as keys to their mental buoyancy under stress.

When the researcher Linda D. Henman interviewed fifty of these former POWs in 2001, though, she discovered another common thread behind their resilience, and it's something that provides us with one final tool for managing stress: humor. The POWs didn't just enjoy having the occasional yuk to break up the tension—they deemed humor a vital ingredient in their very survival. As Henman wrote in one paper on the POWs, "Several of the study participants commented that they considered humor so important that they would literally risk torture to tell a joke through the walls to another prisoner who needed to be cheered up." Throughout the lowest depths of their imprisonment, the Vietnam POWs showed a remarkable eagerness to thumb their noses at their grim situation. The prisoners sarcastically named each building within the complex after a hotel on the Las Vegas strip, like "Stardust," "Golden Nugget," and "Thunderbird." They also often joked with each other that "you have to get here early to get the good deals." McCain still likes to poke fun at the plane crash that shattered his body just before his capture, saying, "I stopped a missile with my aircraft."

Some POWs went much further, as Henman's research reveals. One day in the midst of his five-year incarceration, Colonel Gerald Venanzi spotted a few fellow prisoners tied up uncomfortably near his open cell, and he quickly hatched a plan to buck them up.

Revving up the motor of an imaginary engine, Venanzi pretended to ride a motorcycle around the prison compound, producing the requisite sound effects with his mouth and feigning the occasional wreck. As the guards exchanged uneasy glances, wondering if they had a lunatic on their hands, Venanzi's comrades cracked up at his jest. When his captors sent him to solitary confinement and "took away" his motorcycle (it wasn't fair for him to have a motorcycle if the other captives couldn't too, a commander explained), Venanzi upped the ante. In his isolation cell, Venanzi invented an imaginary chimpanzee friend named Barney Google, and the make-believe primate soon became a fixture at the prison complex. Barney liked to accompany Venanzi to interrogations, wherein Venanzi staged debates with the animal—"I can't tell them that! They'll beat the hell out of me!"—as his questioners looked on in shock. (Once, the camp commander tried to mollify the discord by offering Barney tea, but Venanzi had to relay a polite no; Barney didn't like tea.) Venanzi's Barney Google stories provided limitless amusement to the Americans. Ultimately, the guards summoned Venanzi once more and told him he'd have to release his chimp into the wild: he was getting new roommates, and Barney's presence might upset them.

In tense moments, explains the clinical psychologist Rod Martin, the purpose of pranks like Venanzi's isn't merely to elicit a chuckle; joking actually reformats your perception of a stressor. "Humor is about playing with ideas and concepts," said Martin, who teaches at the University of Western Ontario. "So whenever we see something as funny, we're looking at it from a different perspective. When people are trapped in a stressful situation and feeling overwhelmed, they're stuck in one way of thinking: *This is terrible. I've got to get out of here.* But if you can take a humorous perspective, then by definition you're looking at it differently— you're breaking out of that rigid mind-set." Look at the POWs' decision to rename the camp buildings after glitzy casinos: these were structures where they endured *torture*, yet instead of sticking with a grim perspective on these disturbing places, they flipped their experience over and found a different, ironically amusing way of looking at it. This kind of humor, Martin says, helps us mentally defuse (and even rise above) a source of fear or stress. By poking fun at a stressor, we take away its psychological venom; suddenly, things seem less dire, and our attitude shifts from dread and tension toward challenge and control.

If we wanted to get clinical about it, we could explore how humor activates the brain's ventromedial prefrontal cortex, the very same region that inhibits the amygdala during fear extinction. But at least in this case, common sense outweighs the neuroscience; we intuitively know the value of humor in trying times. We see it in soldiers like Venanzi. We see it in the Londoners who coolly weathered the Blitz, jokingly telling each other, "Nice day for the Blitzkrieg." We even see it in those who lived through horrific stints in concentration camps. As the psychiatrist and Holocaust survivor Viktor Frankl wrote of his efforts to survive Nazi imprisonment, "Humor was another of the soul's weapons in the fight for self-preservation."

And we especially see humor—of the dark, gallows-oriented variety—in those of us who must face the macabre on a daily basis in professional life. People like the veteran Modesto, California, paramedic Rod Brouhard, a man for whom the most horrifying sights are just part of a day's work. "We have this sort of repetitive bad luck whenever I train a new paramedic," Brouhard told me. "Every time I get a new paramedic intern, it just so happens that on that first day, we'll have a dead baby. So my partner and I, we have to joke about it." It's not that wisecracks about dead kids are especially hilarious to Brouhard, who has five young children of his own; it's just that any paramedic who can't find a way to poke fun at these grisly things will break under the psychological strain. According to one 2005 survey of more than six hundred paramedics, 90 percent of them use dark humor to deflect psychic stress. Other research has found that paramedics' use of humor increases as they gain more experience, and that those who *don't* find a way to mock the grimness of what they see become far more likely to burn out. "People who take themselves too seriously in this job, those are the ones you worry about," Brouhard said.

Paramedic humor, I hasten to point out, is so painfully unfunny that in a solid hour of talking to Brouhard, I couldn't locate anything that would make a normal person so much as chuckle. Mostly, it makes you want to go take a hot shower and then watch the Disney Channel, because the things they most need to joke about are also the things that are most disturbing. "I've seen a few train-versus-person cases," Brouhard recalled, warming up some paramedic gag material. "The trains don't even notice you. It's kind of like running over a bug. You're—" You know, let me just summarize Brouhard's next few comic sentences, so you don't have to lurch for an airsickness bag: *Speeding trains are unkind to the structural integrity of the human body.* "You see that kind of stuff and you just…wow," Brouhard continued. "You *have* to joke about it. It's automatic." As a rule, he says, paramedics hold their tongues when someone is hurt or when bystanders are in earshot, but not at, say, Pizza Hut after the call is over. "I have cleared out tables," Brouhard told me. "There were things left uneaten."

Of course, dead children and train-track deaths are tragic. The Holocaust was tragic. Suffering a long stay at the Hanoi Hilton was tragic. But as those like Venanzi and Brouhard have found, there's usually *something* to laugh at in tough times—and if you can do that, you lift yourself above your stress. We don't need to be natural comedians to make use of this strategy, and our quandary doesn't need to strike us right away as a thigh-slapping laugh riot. We just have to make an effort to play around with our negative perceptions of a stressful situation, which, as Martin explains, allows us to see our predicament from a less threatening viewpoint. "In humor, there's kind of a refusal to accept fate," he told me. "When you poke fun at a stressor, it becomes not something that you have to run away from, but something you can handle."

So whether or not we emerged into the world as stress-proof superheroes, we can all make strides in adapting our attitude about stressors. Even if we don't thrive on uncertainty and chaos like Farley's Big T's, we can work at becoming hardier, learning to embrace change and accept the uncontrollable bit by bit. Even if we don't come prepackaged with expert decision-making knowledge in stressful situations, we can train diligently, as Ericsson and Klein suggest, until we better hone our instincts. Even if we can't magically boost the brain's production of neuropeptide Y to Green Beret levels, we can learn to reframe a stressful situation as a challenge and an opportunity to grow. We can stay engaged instead of giving up. We can take small, concrete steps toward our goals, to show ourselves that we are authors of our own fate—not helpless in the face of adversity.

And if all of that claptrap fails, we can always invent an imaginary chimpanzee pal and make fun of the whole mess. But please remember never to serve him tea. That would just be embarrassing.

PART TWO

The Elements of Cool

Thinking Under Fire

Poise in the Spotlight

Clutch Performance

Composure in Crisis

CHAPTER FOUR

Think Fast: Cognition Under Pressure, and How to Improve It

Perched uncomfortably on the hot seat of the game show *Who Wants to Be a Millionaire?*, surrounded by cameras and rapt spectators, Ogi Ogas scrunched his eyes shut yet again, as if to wring the final answer from his brain by force. After fourteen questions, four lifelines, and several miniature nervous breakdowns, Ogas finally stood atop *Millionaire*'s nerve-racking summit; as Meredith Vieira, the show's host, had reminded him so dramatically a moment before, "One question separates you from a million dollars." The ominous heartbeat thump of the show's sound track floated through the tense air of the studio. Ogas squirmed in his chair. Once more, he opened his eyes and stared at the question on the flatscreen monitor before him:

Which of these ships was not one of the three taken over by colonists during the Boston Tea Party?
A: Eleanor   B: Dartmouth
C: Beaver   D: William

Ogas *knew* the answer was lurking somewhere in his brain. He'd been living in the Boston area for a decade, for God's sake. But it wasn't just Ogas's mailing address that made him believe he could figure out the answer. He also knew he had a secret edge.

Ogas, you see, was no ordinary *Millionaire* contestant. When it came to general trivia knowledge, Ogas considered himself a lightweight (he didn't even own a television), yet the hidden engine behind his ascent to the million-dollar question more than compensated for this deficit. A trim, intense-looking man with receding brown hair and black oval glasses (he could be Apple CEO Steve Jobs's younger brother), Ogas was a graduate student in cognitive neuroscience at Boston University whose doctoral research focused on human memory—an especially useful field of expertise for a high-stakes trivia contest. "Basically, what I do is look at how the brain works and try to create computer models or algorithms to model that," Ogas explained. Months before his summer 2006 appearance on the show, Ogas had been contemplating ways to make quick money when he realized he could use his brain science skills to beat *Millionaire*, landing himself an "easy" seven-figure check. "I knew all of these brain techniques," he recalled, "so I thought, *You know, this is a problem I can crack.*" For months, Ogas honed a set of cognitive tricks to help him locate tough answers in the hidden subnetworks of his mind. He mapped out the most effective ways to employ the lifelines, like "Ask the Audience" and "Phone a Friend." He reviewed past *Millionaire* questions, carefully testing his techniques and hunches. Then, that July, Ogas took a bus to Manhattan and aced an audition test at the ABC commissary. Within a few weeks, he found himself in the hot seat opposite Vieira.

Ogas now had just one hurdle left to clear before reaching the million-dollar prize. *Eleanor, Dartmouth, Beaver,* or *William*— which was the odd one out at the Boston Tea Party? With no lifelines left, he'd have to figure it out on his own. "I've definitely read this at some point," Ogas murmured as Vieira looked on placidly. A strong intuition told him that *Dartmouth* was one of the Tea Party ships. He launched into a memory technique called "priming," trying to use the fragment he knew to get his brain to spit out the full answer: "*Dartmouth . . . Dartmouth . . . Dartmouth . . . Beaver.*" The second name flared in his mind like a neural firework. He *knew* it was right. "That voice is talking in your head again, isn't it?" Vieira asked. Ogas continued priming, now imagining a famous painting he'd seen of the Tea Party. "*Dartmouth, Beaver . . . Beaver, Dartmouth . . . Dartmouth, Beaver . . .*" A moment later, in another luminous flash, he had it. "I think it was *Dartmouth, Beaver,* and *Eleanor,*" he said, grinning weakly. Which meant the right answer must have been "*William.*" He'd done it. For one glorious split second, Ogas believed he was about to become a millionaire.

But then something happened that Ogas hadn't prepared for: the pressure started to unnerve him. He considered the stakes. A correct answer would net him $1 million, but if he was *wrong*, the $500,000 that he'd already won (and which he could walk away with at any time) would dwindle to a measly $25,000; he'd risk losing $475,000. "I'm a student, so that's more money than I've made in my whole life," Ogas explained. Fear welled up through his body. Ogas bounced and rocked in the hot seat, suddenly aware of tiny, intimidating details in the set around him—the hum of the audience, the soft powder of Vieira's makeup. "From a purely physical point of view, they intentionally design the *Millionaire* set to be anxiety-producing," Ogas told me. "For example, there's the heartbeat music going *bumpbumpbumpbump,* and they put you on this high, wobbly chair, so you have the feeling that you might fall off at any second—plus, the floor below is glass, which adds to this effect of being on a flagpole. And, of course, the audience is surrounding you three hundred and sixty degrees, like you're in a coliseum." He thought of his wife sitting in the audience, her hands clasped under her chin, and how he promised her he wouldn't gamble away a fortune on intuition alone. "I remember waves of adrenaline pouring over me, and I just couldn't pull the trigger," Ogas recalled. "I froze." Shaking his head in resignation, he sighed to Vieira, "Ah, damn it. I can't. There's too much at stake. I'm going to walk." He'd keep the half million. The audience applauded his prudence.

As Ogas describes it today from the bitterly comfortable confines of the downtown Boston condo he bought with his *Millionaire* winnings, the next fifteen seconds felt roughly as though several invisible wild dogs were playing tug-of-war with his entrails. "Ogi, let's see the answer, so you know," Vieira said, and as "*William*" flashed in green on the monitor, Ogas exploded out of his chair so violently that it seemed like he'd been nudged with a cattle prod. "*Damn it!*" he screamed, flailing his arms in fury. For a moment, Vieira appeared to be contemplating calling in security to restrain him. ("Now that's good television!" she exclaimed after the tantrum passed.) Even as Ogas walked over to hug his wife, the seeds of what was to come were already visible on his face. Despite his considerable payout, he couldn't forgive himself for cracking under the pressure. "I was just in a terrible depression for the next few months," Ogas told me, "and the main source of it was that I had trained and prepared and all of my techniques worked—even with the million-dollar question—but I lost faith because of all of that adrenaline. The stress and the anxiety got to me. I was angry with myself. I *should* have prepared for that pressure." Ogas begged the universe for a chance to redeem himself.

Less than a year later, he got that chance. In June 2007, Ogas received a phone call inviting him to participate in a trivia battle royale called *Grand Slam*, which would pit sixteen legendary game show champions against one another in a winner-take-all tournament for $100,000. The show, scheduled to air on the Game Show Network with Dennis Miller as its host, posed a daunting challenge. Each episode featured two contestants clashing head-to-head in four lightning-fast trivia rounds, competing to answer a chain of questions as quickly as possible while a digital timer ticked away above their heads. And these opponents were no slouches — they were highly decorated game show masters, veterans of juggernauts like *Jeopardy!*, *Twenty-One*, and *Millionaire*. Among them stood two trivia titans who possessed, in Ogas's estimation, virtually superhuman memory skills: Ken Jennings, the celebrated *Jeopardy!* winner who became a household name during his run of seventy-four consecutive victories, and Brad Rutter, the all-time *Jeopardy!* champion, with more than $3.2 million in winnings.

Right then, Ogas resolved to bury the disgrace of his *Millionaire* implosion under a mountain of *Grand Slam* glory. This time around, he would bring the full cognitive-science arsenal to bear, and send tremors through the game show firmament. "Brad Rutter and Ken Jennings, they're famously smart, viciously smart," Ogas told me. "And I said, 'You know what? I'm going to win this. I'm going to beat the shit out of every one of them. And the key to it is that I'm going to find some way to beat the pressure — to be cool as a cucumber under that stress. I'm going to beat this thing, and this time I'm going to beat it emotionally, too.'" Ogi Ogas pledged to become a master of thinking fast under pressure.

Beware of Heat-Ray

It's never been a secret that we find it tough to think clearly in stressful, anxiety-provoking situations. Writers and thinkers have harped on this idea for centuries. The legendary Anglo-Irish intellectual Edmund Burke encapsulated it nicely in 1757, when he wrote, "No passion so effectively robs the mind of all its powers of acting and reasoning as fear." Bertrand Russell, the great British philosopher, likewise remarked, "Neither a man nor a crowd nor a nation can be trusted to act humanely or to think sanely under the influence of a great fear." Even Frank Herbert, in his science-fiction opus *Dune*, famously noted that "fear is the mind-killer."* Put simply, our brains just aren't designed to think and reason at top speed when the amygdala's guns are blazing. As we've already learned, the amygdala actually overrides the higher cognitive functions of the cortex as it revs up, sending us fighting or fleeing in a snap—which works spectacularly on the lion-infested Serengeti, but not nearly as well in the *Millionaire* hot seat. In the last chapter, we saw how hard-won intuition enables experts to make decisions under fire without thinking, yet when people *do* have to deliberate and problem-solve while afraid, pressured, or stressed...well, the results can be mixed.

Before we delve into the ways that fear can cloud our ability to think clearly, a vital caveat is in order: fear doesn't *always* impair rationality—in fact, it carries plenty of cognitive *benefits*. We need fear not just to help us flee from danger, neuroscientists say; we need a modest dose of it to think wisely, as well. Psychologists first discovered this advantageous side of fear back in 1908, when experiments by Robert Yerkes and John Dodson found that human performance peaks when we hit a certain optimal level of nervous activation. If we're not anxious at all, they found, we approach tasks with a disinterest that is as toxic to performance as flat-out terror. While different tasks have different optimal anxi-

---

* Please ignore that Herbert's character reflects on this idea as a strange old woman forces his hand into a mystical green cube while holding a poisonous weapon called a gom jabbar to his neck.

ety levels—for instance, strenuous physical tasks get easier with more adrenaline—the one constant that Yerkes and Dodson saw is that we always need *some* fear and anxiety to do our best. (We see this especially in those who feel they do their best work up against a deadline; the extra shot of anxiety motivates and focuses them.) This insight also applies when we're trying to achieve first-rate cognition. Contrary to the widely held belief that a perfect decision-maker would ponder each choice on a Spock-like emotionless plane of pure reason, research has shown that we simply can't make good decisions without fear. The neuroscientist Antonio Damasio has performed several studies of patients with major amygdala damage—patients who essentially can't feel fear—and these people often struggle to make the simplest of choices, like which peanut butter to buy at the grocery store. Fear is a crucial link in the cognitive chain; behind the scenes, it helps nudge us in the right direction.

But the emotion also has its intellectual pitfalls, and for the prime historical case of fear bulldozing the mind's powers of reason, we now turn our gaze to a quiet autumn night in 1938, when 6 million Americans tuned their radios to the Columbia Broadcasting System for a relaxing hour of terror-inducing, hysteria-igniting apocalyptic entertainment. On October 30, the day before Halloween, Orson Welles and the *Mercury Theatre on the Air* ensemble decided to celebrate the upcoming holiday in style, with a spooky adaptation of the Martian-invasion masterpiece *The War of the Worlds*. Welles wasn't necessarily aiming to be provocative. He and his co-writers figured that the story, drawn from the 1898 H. G. Wells novel of the same name, was so well known and so outlandish that no one could possibly interpret it as anything other than a diverting radio drama. They were wrong.

The problem was that the *Mercury Theatre*'s radio play was a little *too* good. From a dramatic standpoint, the 1938 *War of the*

Worlds broadcast was a work of bloodcurdling brilliance; even seventy years on, it still gives you chills. After a brief introduction (which makes clear that listeners were about to hear a work of fiction), the radio play opens in the most mundane way possible: an announcer gives a weather report and then sweeps the audience to the Meridian Room at New York's Hotel Park Plaza for a musical performance by "Ramon Raquello and His Orchestra." To those who tuned in here, as most apparently did, CBS's offering for the evening truly seemed to be this platter of light orchestral pieces. Soon enough, another news announcer cuts into the music for a special bulletin, informing listeners that "several explosions of incandescent gas," each looking "like a jet of blue flame shot from a gun," have just been observed on Mars. No big deal. The broadcast returns to Raquello for a rendition of "Stardust." Over the next few minutes, the announcer steps in again and again with increasingly distressing updates. A Princeton professor, played by Welles, speculates about the possibility of life on Mars. More music. A suspiciously smooth-looking meteorite plummets onto a farm in Grover's Mill, New Jersey. How about a swing tune? To the unsuspecting listener, it all sounds perfectly credible.

Now the cataclysm begins to unfold. Carl Phillips, a reporter on the scene in Grover's Mill, narrates as the meteorite emits an eerie hum, then slowly unscrews itself. Tentacles emerge, followed by a dark creature as "large as a bear" that "glistens like wet leather," with "luminous disks" for eyes and "rimless lips that seem to quiver and pulsate." It's all downhill from here. The alien fires "heat-rays" at the crowd, incinerating them instantly. Fleets of Martians mount giant tripods of doom, spewing poisonous gas as they overtake Manhattan. Military generals and government officials tell of their futile attempts to thwart the aliens. Within a tidy forty minutes or so, the New York metropolitan area has become a desolate wasteland, with a lone radio operator

calling out, "Isn't there anyone on the air? Isn't there anyone..."
Silence.

For the majority of the program's listeners, this came across as a cracking radio drama. For the rest, it came across as the end of life as we know it. Out of the 6 million Americans who tuned in, somewhere between 500,000 and 1 million horrified people believed Welles's radio play was a genuine news broadcast depicting the destruction of human civilization. Thousands of listeners packed their families into cars and raced west, away from the Martian attack site. As the *New York Times* reported the following day, "In Newark, in a single block at Heddon Terrace and Hawthorne Avenue, more than twenty families rushed out of their houses with wet handkerchiefs and towels over their faces to flee from what they believed was to be a gas raid." (The *Times* switchboard also received nearly a thousand calls from listeners wondering where they'd be safest from heat-ray attacks.) Others took a more proactive approach: several reports emerged of well-armed citizens opening fire on water towers, mistaking them in the darkness for Martian tripods. After the *Mercury Theatre* producers fielded calls from panicked listeners, Welles broke in several times to remind the audience that they were performing a radio play, but the damage had been done. In the following days, Welles took a merciless beating in the press, with indignant citizens and grandstanding politicians lining up to condemn the stunt. The commissioner of the Federal Communications Commission went so far as to label Welles and his company "terrorists."

What you *didn't* see much of in the tense aftermath, however, was the asking of a fundamental question: how could so many people fail to apply a shred of common sense to what they were hearing? While the *War of the Worlds* furor was still fresh, the Princeton psychologist Hadley Cantril began investigating this very issue, and his findings reveal an intriguing portrait of how fear can distort the human mind. After all, it wouldn't have taken much effort for listeners to figure out that the whole thing was staged; they might have asked, "Isn't it odd that CBS already had microphones set up with these field artillery units?" or "The reporters seem oddly eloquent and composed as they're being annihilated by the Martian horde, don't they?" Yet Cantril found that between a quarter and a third of all listeners made *no attempt whatsoever* to confirm or refute what they were hearing; they didn't even switch to another radio station or check with their neighbors. Many in the audience were so fixated on their terrified interpretation that they had no recollection of Welles's disclaimers and reminders, even though they had listened straight through them. Some in the New York area hallucinated that they could smell poison gas or see the smoke from the heat-ray attacks looming above Manhattan. For most, the single element that was most crucial in convincing them of the broadcast's authenticity was the appearance of so-called experts like Princeton professors and government officials. Under the sway of fear, these people became alarmingly gullible marks, their capacity for reason placed on temporary standby.

We might like to think of events like this as bizarre anomalies in the annals of human psychology, but similar fear-based thinking errors happen to us all the time. For example, consider how we analyze the hazards in our lives — or, as the University of Oregon psychologist Paul Slovic might put it, how astonishingly bad we are at knowing what we should be afraid of. For nearly fifty years, Slovic has been studying how the human mind calculates risk, and his work has uncovered dozens of hidden biases that lead us to miscalculate how dangerous things really are, from handguns to global warming to nuclear power. We generally assess risk not through rational analysis, Slovic says, but through our emotions, with fear taking a starring role. "For example, if a potential outcome is vivid in our minds and triggers a strong negative feeling, we treat it as if it were the likely outcome," Slovic told me. "Even a very low probability event, like a murder, creates such a strong negative sense that it confuses us into thinking that it must be likelier than it is. We react just as strongly when the probability is low as we do when it's high." In other words, we get fooled by vividness, which makes us vastly overestimate the likelihood of the kind of fearsome events we see on the nightly news—fire, homicide, kidnappings—and react accordingly. Driven by an alarming mental image of a hooded intruder breaking into our house, we go buy a handgun for protection, foolishly believing we're somehow exempt from the statistics showing that a gun discharged inside the home is *twenty-two times* more likely to hurt a friend or family member than a burglar or assailant. Fear trumps logic.

For Slovic and the risk-assessment crowd, there is no better illustration of dramatic emotional images clouding our rationality than America's collective response to the threat of terrorism. The September 11 attacks terrified hundreds of millions of Americans, and with good reason; the footage of the planes hurtling into the Twin Towers is some of the most disturbing imagery ever captured on video. Yet there's also cause to believe that the psychological effect of the 9/11 attacks was just as deadly as the act itself. After 9/11, for instance, many frightened Americans began to view air travel as exceedingly dangerous. So, whenever possible, they took to the roads instead—a decision that made *emotional* sense, but not *rational* sense. As one MIT professor has determined, every time we step on board an airplane for a domestic commercial flight, our odds of dying are approximately one in sixty million. Driving the distance of a typical flight, on the other hand, carries a risk of fatality roughly sixty-five times greater than airline travel. And what happens when more people choose driving over flying? More people die. In 2008, a team of Cornell University researchers analyzed traffic statistics from the period after 9/11, and after controlling for variables like weather and road conditions, they calculated that the mass shift toward car travel resulted in as many as 2,300 driving fatalities that wouldn't have happened if people had stuck to their old flying habits. By contrast, the 9/11 attacks had an estimated death toll of just over 3,000. Thus, in a very real sense, *fear of terrorism* has proven nearly as lethal in the United States as terrorism itself.*

Meanwhile, as we busy ourselves overreacting to vivid threats, we *under*react to hazards that unfold slowly, chronically, and out of sight. Take humankind's tepid reaction to global warming, the potentially world-altering future calamity that makes most people yawn with boredom. "Global warming, to use a terrible pun, doesn't hit any of our hot buttons," Slovic explained. "Even though it could be very serious, it's not the kind of thing that gets us motivated. It's distant, it's hard to imagine, and it's natural. We tend to underreact to nature. We think it's benign and familiar, so we don't take nature's risks as seriously as we should." It's the same story with health hazards: we drive ourselves to hysteria over immediate, catastrophic ailments like avian flu and mad cow disease (neither of which has had a substantial impact in the United States to date), while remaining largely indifferent to slow-building, chronic conditions like diabetes and heart disease (which kill 72,000 and 630,000 Americans each year, respectively, according to 2006 CDC statistics). We can bring our powers of logic to bear on these risks if we really try, Slovic says, but we typically don't. Right or wrong,

---

*Just for the record, according to a *New York Times* report, statisticians have concluded that the average American's odds of perishing in an act of international terrorism on U.S. soil are "comparable to the risk of dying from eating peanuts, being struck by an asteroid, or drowning in a toilet."

our beguiling emotional intuitions like to seize the psychological limelight, leaving reason to look on from the sidelines.

We could go on for hours about the ways fear can hinder cognition—it makes us form harebrained superstitions; it makes us see patterns where none exist (*see* stock market, the); it makes us buy into fearmongering political ads, causing us to vote for dimwitted candidates with lantern jaws and windproof hairdos. Suffice it to say that the more fear you feel, the more difficult it is to think rationally, creatively, flexibly, unhurriedly, or realistically. But thinking well under fire is far from impossible.

How Smart People Choke

Grand Slam may well be the only game show in history that could make its viewers have a seizure, purely from the strobe-light speed of the trivia action onscreen. The creators of the program, which originated in the United Kingdom, designed its format with the explicit goal of putting Britain's most decorated quiz show champions through as nerve-racking a cognitive challenge as possible. They succeeded; each episode of *Grand Slam* comes across as a hybrid of high-tension bomb defusion and quick-talking livestock auction.

The show works as follows. Two contestants stand facing each other atop what looks like a glowing white Rubik's Cube, separated by only a couple feet of space. A giant video screen looms behind them, showing close-ups of the competitors' taut faces. Above each challenger's head hovers a white digital timer. The rules of the game are simple: In the first three trivia rounds, the contestants both begin with one minute on their clock, and their objective is to answer questions as rapidly as they can. Each player's timer begins ticking down as soon as he's asked a question, and it freezes again once he gives a correct answer, thereby switching the spot-light over to his opponent. (If a contestant answers incorrectly, he receives another question while his time continues draining.) The player whose clock reaches zero first loses the round, and the winner gets to keep whatever time he has left to use in the decisive final round. With such a premium on speed, the overwhelming effect of *Grand Slam* is of a game show on amphetamines. Here, for instance, is a snippet from the grand finale of the 2003 U.K. version, which featured two pasty gentlemen named Gavin Fuller and Clive Spate. Keep in mind: the characters aren't talking like this; *THEY'RETALKINGLIKETHIS!!!*

ANNOUNCER: Clive, also known as Donar, who is the Norse god of sky and thunder?

CLIVE: Thor!

ANNOUNCER: Correct. Gavin, what is fifty-four minus twenty-nine minus seven?

GAVIN: Eighteen!

ANNOUNCER: Correct. Clive, who was the first ever DJ heard on Radio 1?

CLIVE: Tony Blackburn!

ANNOUNCER: Correct. Gavin, in Scrabble, how many *F* tiles are available?

GAVIN: Two!

ANNOUNCER: Correct. Clive, built by Carl Langhans, what is Berlin's only remaining town gate?

CLIVE: Brandenburg Gate!

ANNOUNCER: Correct. Gavin, what name is given to the short tail of a hare, rabbit, or deer?

GAVIN: Scut!

ANNOUNCER: Correct. Clive, who, in 1921, was the first prime minister to occupy Chequers?

CLIVE: Lloyd George!

ANNOUNCER: Correct. Gavin, what is eight hundred and six divided by thirteen?

GAVIN: Sixty-two!

Which was, naturally, correct. The cerebral flurry above took under thirty seconds to play out.

For Ogi Ogas, this blistering pace posed a serious problem: the cognitive tricks he had developed for his *Millionaire* appearance were based on having unlimited time to answer a set number of questions. On *Millionaire*, every contingency could be carefully planned for. If he needed to "Phone a Friend," for example, Ogas knew that his BU colleague Sai Gaddam would be on the other end of the line with a tailor-made search engine program, specially designed to process multiple-choice *Millionaire* queries. (As will become abundantly clear, one could never accuse Ogas of underpreparing.) Likewise, Ogas trained for the "Ask the Audience" lifeline by studying the work of his academic adviser, Gail Carpenter, whose research showed how he could surmise the answer to a question by seeing how his perspective fit in with the larger crowd. On *Grand Slam*, however, such tricks would be useless.

The severe mental requirements of *Grand Slam* provided Ogas with a thrilling challenge, a new cognitive puzzle for him to solve. Actually, to be more accurate, it provided a puzzle for him to *obsess over*. A man predisposed to the furrowed brow and the ever-churning neurotic mind, Ogas describes himself as "a schemer and a grinder." His colleagues at Boston University speak of his ability to focus on problems with laserlike intensity; as Carpenter once put it, "Ogi Ogas is one of those rare students who learns as if his life depended on it." (On the other hand, Ogas sometimes fails to take notice of certain vital details of the real world. When

I asked him how old he was, Ogas replied, "Uh, thirty-six or thirty-seven...Wait, what year is it?") Ogas's path to cognitive neuroscience has been a circuitous one—before enrolling at BU, he drifted through France, Russia, and even Hollywood, where he took a stab at screenwriting—but as he explains it, his obsessive leanings found their perfect match in "thinking about thinking."

Grand Slam offered Ogas an opportunity to delve into his special passion: the intricate workings of human memory. Competing against a group of game show demigods, Ogas knew he was outgunned in the trivia and brain-wattage departments; the question was, could he overcome this mismatch and turbocharge his memory with the aid of neuroscience? "I realized *Grand Slam* would be an epic battle between two kinds of minds: big brains vs. brain science...Superman vs. Lex Luthor," Ogas wrote me in an e-mail. "Brad Rutter and Ken Jennings are both natural-born supercerebrums, possessing cortical configurations for memory and problem-solving more brawny and potent than those of mere mortals. I, myself, am mortal—an envious and scheming Lex Luthor." Whatever Ogas lacked in cerebral firepower, he'd make up for in sheer effort.

For the two months leading up to his appearance on *Grand Slam*, Ogas dedicated himself to a full-time mental training regimen. In part, this entailed the development of another set of cognitive shortcuts for typical *Grand Slam* questions; he learned, for example, how to multiply two-digit numbers by eleven (add the two digits together, then put the sum between the original digits), how to do lightning-fast long division, and how to unscramble anagrams quickly. This last technique, Ogas's own creation, filled him with particular satisfaction. Take this anagram of a celebrity's name: IRPHLOSATNI. Ogas discovered that because of the way the eyes and brain work together to process text, if you look *above* a word jumble instead of right at it, the brain has an easier time sorting out the answer.*

The lynchpin of Ogas's *Grand Slam* strategy, though, was his cunning, multitiered scheme for memorizing trivia. Essentially, Ogas structured his knowledge banks like the memory system of a computer. The lowest layer of information in this configuration was what Ogas called Tier Three, an array of four thousand obscure factoids that he looked over once and marked for cold storage in his deep memory. Next up came Tier Two, a group Ogas described as "five thousand facts that would probably be asked on *Grand Slam*, but were unfamiliar to me: *Desperate Housewives*, vichyssoise, Justin Timberlake, Brett Favre," and so on. Ogas likened this data to files on a hard drive, and to keep it relatively fresh in his mind, he reviewed it once a week. The topmost layer of information Ogas vacuumed up, Tier One, consisted of 1,500 facts of paramount importance—landmarks, U.S. presidents, historical events—that needed to be at the forefront of his mind at all times, lingering in his short-term memory reserves. Ogas wanted his Tier One knowledge to be like RAM in a computer, which temporarily stores data for instant user access. To get a similar result from his brain, Ogas had to pore over these factoids constantly, even right before the show itself.

For its mind-twisting complexity alone, Ogas's three-pronged strategy for trivia domination would have made any supervillain shed tears of pride. Yet like every plot Lex Luthor ever hatched to defeat Superman, it also suffered from a critical weakness: it hinged on quick access to short-term memory, but recent research has shown that when we're nervous or threatened, this is the kind of memory that is most prone to falling apart. The detrimental

---

*Which is PARIS HILTON.

effect of anxiety on short-term memory is so pronounced, in fact, that it has recently led psychologists to an astonishing discovery. The smarter you are, it turns out, the more susceptible you become to choking under pressure in mental tasks.

Joshua Aronson, an NYU psychology professor who specializes in test anxiety, can explain. In the mid-1990s, when he was a postdoctoral student at Stanford University, Aronson and the psychologist Claude Steele developed an innovative technique for eliciting choking in exam-takers. In a landmark experiment, Aronson and Steele invited two groups of Stanford undergraduates to take a difficult thirty-minute written test composed of verbal questions from the Graduate Record Examination. Each group contained an equal number of black and white students, all of them comparably intelligent. With one group, Aronson and Steele explained that the test was a relatively meaningless "problem-solving" exam and had them plug away; as expected, the black and white students all performed similarly under this condition. But the second group's exam was different. "We just changed the psychological situation of why they thought they were being tested," Aronson recalled. "Now we framed the test as 'This is going to stick a dipstick in your brain and figure out how intelligent you are.'" The paper exam was the same. The students were still equally bright. Yet suddenly, with this slight tweak to the test's introduction, the black students performed dramatically worse than the white students. They choked.

So what happened? Labeling the exam a measure of intelligence, Aronson says, introduced an anxiety-provoking variable for the black students: especially at such a prestigious university, they all felt weighed down by the stereotype that black students are less intelligent than white students. "Basically, it engaged a fear of being evaluated," Aronson told me. "It's such a little twist, but the internal experience of it is huge. It's sort of like flipping a light switch—it takes less than a calorie to do it, but it changes everything." The black students' anxiety about performing as well as the white students brought a wash of mental noise that hindered their ability to focus. "Under this condition, you're trying to divide attention between at least two concerns," Aronson said. "One is answering the problems right. And then another is all of these intrusive thoughts and worries that are future focused: *What if I look dumb? What are the implications of not performing well?* People who aren't burdened by those concerns are thinking about only one thing, which is *How do I do this?*" In any mental pursuit, we all need access to what psychologists call "working memory," a short-term mental scratch pad on which we hold pieces of information and reason out problems. But working memory is a limited resource, and intrusive worries suck up its bandwidth, making us, in effect, less intelligent than we really are.

This is what befell the black students: cueing the negative stereotype about intelligence, however subtly, introduced thoughts that cluttered their working memory. Aronson and Steele dubbed this effect "stereotype threat," but it's an equal-opportunity affliction. It doesn't just hurt groups that suffer discrimination, like women and minorities—it can torpedo white males as well. In one study, Aronson gave an eighteen-question math exam to mixed groups of gifted white and Asian students; the white students who were told that the test was part of an investigation into Asian math supremacy missed an average of three more questions than the white students in the control group.

Far more so than the physical sensation of anxiety, this conflict between intrusive worries and short-term working memory is what leads to choking on mental tasks: it's like trying to think clearly while juggling several extraneous ideas in your head. And as one 2005 study by the psychologist Sian Beilock showed, this is especially bad news for high academic achievers, because "smart" people tend to use more complex working-memory-intensive strategies to figure out problems. In her experiment, Beilock separated college undergraduates into two groups according to their working memory capacity and then gave them a math test. When subjects took the exam with nothing at stake, the "smart" high-working-memory group predictably outscored the low-working-memory group. Yet when Beilock gave both groups the test again, this time adding in the prospect of a cash prize and promising that a panel of math professors would judge their results, something surprising happened: the low-memory group's scores remained unchanged, while the high-memory group's scores dove all the way down to the level of the low-memory group. Being "smarter" made the high-working-memory group more likely to collapse under pressure; their mental scratch pads suddenly filled with performance-related worries, holding them back from their potential. As Beilock and her co-author Thomas Carr explained it in one paper, "Performance pressure harms individuals most qualified to succeed by consuming the working memory capacity that they rely on for their superior performance."* That is, smart people tend to choke.

With students, at least, psychologists like Aronson have devised a number of useful strategies to combat test anxiety and stereotype threat. Training students to focus on the external task at hand instead of on their internal emotions and worries is the most obvious structural fix. Even students who feel intense anxiety about an exam can often make huge performance strides if they shift their attention away from exaggerated fretting about results

---

* For this very reason, Beilock and many others have argued that high-tension tests like the SAT aren't a fair gauge of actual intelligence — they just measure a student's ability to take pressure-packed standardized tests. Some also argue that the SAT should ask its demographic questions at the *end* of the test rather than at the beginning, to avoid cueing up threat-inducing racial or gender stereotypes before the exam gets under way.

and back to the concrete question before them. One way to facilitate this is to find subtle ways to show kids that their test scores don't define them. "This kind of affirmation says, 'You're going to be seen here not just in terms of your intelligence, but as an entire person,'" Aronson told me. "Studies show that if you give kids a chance to write or talk about things unrelated to academic achievement—their sense of humor, or their ability in sports or music—then they go on to get better grades." Another effective method is to demonstrate to kids that intelligence is a malleable skill that can be improved with effort. Aronson's research has revealed that kids who are cued to believe they can improve their test scores with practice tend to see exams as a challenge rather than a threat, performing better—and feeling less anxiety—than those who see their brainpower as set in stone. "It leads kids to see themselves as works of art in progress," Aronson explained. "I love that idea."

For Ogas, however, affirmations and theories of intelligence weren't going to be enough to keep his instant-access working-memory system humming at its peak level on *Grand Slam*. After his *Millionaire* meltdown, he knew what it felt like to be psychologically unmanned by tension. "If you look at the clip of me answering the million-dollar question, I'm visibly nervous, I'm bouncing in my chair, I'm pulling my hair," Ogas told me. "That intense emotionality impaired my ability to make decisions." He wasn't going to let that happen again. This time around, Ogas would address his anxiety directly—not by trying to banish it, but by working *with* it.

This Is Your Brain on Fire

Broadly speaking, there's a good way and a bad way to think and problem-solve under pressure. To illustrate this, let's take a look inside two very different airplane cockpits.

We'll start with the bad way. Late on the moonless evening of December 29, 1972, as Eastern Air Lines Flight 401 was making its final descent into Miami International Airport, the plane's crew noticed that a green indicator light on the control panel, designed to show that the front landing gear had come down properly, wasn't illuminating. This meant one of two things: either the bulb had burned out, or the landing gear mechanism had malfunctioned—in which case someone would have to crank it down manually. The veteran captain Robert Loft set a westward course over the pitch-black Everglades, and once he reached a cruising altitude of two thousand feet, he put the plane on autopilot so his three-man cockpit crew could investigate their dilemma. One crew member climbed down into the avionics bay to peek at the landing gear. The other two fixated on the indicator mechanism in the cockpit. Those readers who are math experts will have figured out by now that this left no one paying attention to where the plane was going. Thus, no one noticed when one of the crew accidentally nudged the steering column, deactivating the autopilot and putting the plane into an imperceptibly gradual descent. And no one heard the chime of the altitude warning in the engineer's station. The crew was so consumed with fixing the malfunctioning light that they didn't become aware of their change in altitude until they were literally a few feet over the surface of the swamp. By then, it was too late. With the aircraft still hurtling along at 227 miles per hour, its left wing plowed into the water, launching the plane into a violent, flaming somersault. Of the 176 passengers on board, only 75 survived the crash.

Sadly, this wasn't an isolated incident. In fact, a similar accident happened in 1978 only a few miles from where I live, in Portland, Oregon. As the cockpit crew of United Airlines Flight 173 obsessed over another landing-gear-light issue while engaged in an hour-long holding pattern near the city, their Douglas DC-8 simply ran out of fuel and dropped from the sky. No one had been monitoring the fuel gauge. This tendency to become so absorbed in solving an immediate crisis that you forget to actually *fly the plane* is common enough (particularly in student pilots) that military aviators have a nickname for it: "helmet fire." In a stressful predicament, fear can make a pilot's attention narrow in on the danger at hand, ignoring the surrounding environment. He loses situational awareness. He grows fixated and inflexible. He becomes frantic and impulsive, responding — in one psychologist's words — "with great alacrity in poorly planned ways." You can practically see the smoke billowing out from his ears.* Helmet fire is an easy trap to fall into, but it's a terrible (and occasionally disastrous) way to solve a problem under fire.

For a better approach, let's now look in on United Airlines Flight 232, a plane that took off routinely from Denver's Stapleton International Airport on the afternoon of July 19, 1989, and landed again two hours later in one of the legendary feats of modern aviation. At 3:16 p.m. central time, as Captain Al Haynes guided the DC-10 and its lunching passengers through the clear blue sky above the cornfields of Iowa, a thunderous blast suddenly rocked the aircraft. At first, Haynes, who had flown for United for more than thirty years, assumed his plane had suffered an act of terrorism. "I'd never heard anything like that in flight before, just a huge explosion," Haynes told me from his home in Seattle. "It was so loud, I thought it was a bomb." The truth turned out to be even more vexing. Haynes and his crew quickly determined that they hadn't been bombed, so they assumed instead that they'd somehow

---

*A similar problem, called "brain lock," sometimes affects skydivers when their main parachute fails to deploy; they focus so intensely on fixing it that they actually forget to pull their reserve chute, plummeting to their death.

lost one of the jet's three engines. Employing his best unruffled pilot voice, Haynes popped on to the intercom to reassure his passengers that everything was okay—the plane could fly just fine on two engines. He hung up, then began preparing to shut down the malfunctioning tail engine. And it was right about then that First Officer William Records told Haynes something shocking. "Al," he said, "I can't control the airplane."

Haynes swiveled around to look at Records, and what he saw was so bizarre that it almost didn't compute: Records had pulled the control yoke all the way back and to the left, yet the aircraft was descending and banking right. It was a vision that bordered on the impossible. Though the flight crew didn't know it at the time, their plane had just experienced a malfunction too outlandish even to merit a lesson in flight school: the explosion in the tail engine, caused by a failed fan disk, had ruptured the hydraulic lines that powered *all* of the aircraft's control surfaces—like the tail rudder and wing ailerons—leaving them unable to steer the plane. "We had absolutely no flight controls," Haynes explained. "The odds they later gave for this kind of accident were ten to the minus-ninth power—one in a billion. It's just not supposed to happen." This placed Haynes and his copilots in an unimaginably terrible predicament. In the history of aviation, there was no protocol for handling the kind of problem they now faced. Seven miles up in the sky on an out-of-control airplane, with nearly three hundred people on board and no idea how to save them, the flight crew had to figure out a way to land safely, and do it fast.

From the moment their ordeal began, the United 232 crew acted with flawless poise. "We reacted without thinking about the seriousness of the situation or the consequences," Haynes said. "You just have to take care of the problem. Keep this thing right side up, keep it going until we're on the ground—this is your thought process." Because of the blast damage to the aircraft's tail, the plane wanted desperately to pull right and roll over; their first task was to get it flying straight again. The crew soon discovered that they could still control the throttles for the engines on each wing (they were pretty much the *only* thing they could control), so Haynes immediately jammed the left throttle shut and opened up the right throttle, which brought the plane level. Right away, Haynes requested ideas for what to do next. "Working together was what kept us going," he said. "If somebody came up with a suggestion, we'd try it. Some captains just wouldn't take advice, but if that had been the case in our situation, I think we'd have crashed — because this captain didn't know what the heck was going on."

After much trial and error, Haynes, his copilots, and a flight-instructor passenger (who came up to the cockpit to volunteer his services) developed a method for maneuvering the plane using only the two throttles, a process Haynes has likened to steering a car down a steep mountain slope just by opening and shutting its doors. Explained Haynes, "It was a matter of moving the throttles in reaction to what the plane was doing. If you wanted to turn left, you just put more power into the right engine. When the nose wanted to go down, we had to push both throttles up to lift it." In this wobbly way, United 232 beat an unsteady path toward the Sioux City airport. Amazingly, Haynes even showed a flash of humor mere seconds before what promised to be a rocky landing. When the Sioux City air traffic controller informed Haynes that United 232 was "cleared to land on any runway," Haynes chuckled and replied, "You want to be particular and make it a runway, huh?"

Haynes's words were all too prescient; the flight crew's cool couldn't cancel out the whims of fate. Just as the DC-10 was gliding onto the runway, its nose abruptly pitched downward — and the crew was powerless to stop it. The aircraft's right wing struck the tarmac first, spewing fuel and dragging across the runway like a match against sandpaper. A plume of flames erupted from the jet as its nose hammered into the ground. With a crack, the cockpit snapped off from the fuselage and spun into a nearby cornfield, knocking Haynes unconscious. The crash was so violent that passengers found themselves unable to stop their arms and legs from flailing wildly about. Yet the majority of those on board United 232 survived: out of the 296 passengers and crew on the aircraft, 185 lived. Squeezed by life-or-death pressure, Haynes and his crew thought out an answer to a dilemma many pilots would have considered hopeless. In fact, when a variety of DC-10 pilots tried to replicate their feat on flight simulators afterward—with no one's life on the line and plenty of time to plan their strategy— they all failed, over and over again. Haynes's crew had innovated a one-in-a-million solution to a one-in-a-billion problem, and they had done so in the shadow of intense fear.

These are the two extremes of thinking under pressure: the Al Haynes way and the helmet fire way. One is open and flexible, while the other is narrow and rigid. One is aware of what's happening in the surrounding world, while the other is fixated only on the immediate threat. One is focused and realistic, while the other is frantic and rash. And, crucially, one works, while the other doesn't. Fear was a constant in both cockpits—and how could it not be? But what truly mattered was how the two captains channeled it.

When Ogi Ogas faltered on his final *Millionaire* question, he experienced a classic helmet fire moment. His thinking, so much clearer when the stakes were low, grew frenetic and cluttered, dominated by worry. "My mind absolutely narrowed in on one option," Ogas recalled. "Even though I had the answer, I just couldn't let go of the idea of walking away. I was so terrified that I couldn't snap myself out of it." On *Grand Slam*, however, Ogas wanted to think more like Al Haynes: clearheaded, aware, task-focused, flexible. With this aim in mind, Ogas set out to learn how to think while afraid—and that's an important distinction. Like Haynes in the DC-10 cockpit, Ogas never tried *not to be afraid*, a pursuit that would have been as unnecessary as it was foolhardy. High fear and anxiety can make it difficult to think clearly, but as we saw earlier, too little fear is every bit as harmful to cognition as too much of it. Ogas *needed* fear.

Thus, the goal of his emotional training was learning not how to banish fear but how to direct it. To use his words, he sought to become an "amygdala whisperer." Said Ogas, "I knew I would need to be able to do cognitive calculations under stress on *Grand Slam*—so I trained under stress." Adhering to his usual strategy of obsessive overpreparation, Ogas threw himself into an emotional boot camp, putting his brain through trial runs under conditions of high anxiety. To gear up for the stress of performing in front of an audience, Ogas strolled to various crowded public places in Boston and simply started talking out loud. At Boston Common, he recited the names of the presidents backwards. Inside the Downtown Crossing Macy's store, he enlightened customers about square roots. By design, the experience horrified him as much as it did the innocent bystanders who just wanted to pick up a pair of socks, not learn the history of France from a strange babbling man. "I got looks, but that was the whole idea," Ogas recalled. "I never do anything that puts me in public like that, so the first few times there were a lot of distracting emotions. But after I'd done it a few times, I just became acclimated."

As with his groundwork for *Millionaire*, much of Ogas's emotional regimen for *Grand Slam* was rooted in neuroscience. Whenever he was cramming his brain with trivia, for instance, Ogas liked to jack himself up with enough caffeine to give a rhinoceros the jitters. This practice had a specific purpose: to get his brain used to thinking while nervous. "The idea behind that is something called state-dependent memory," he explained. "It's easiest to recall something when you're in the same mental state as when you first learned it. So I drank coffee to get my emotions all riled up because I specifically wanted the ability to recall information quickly under those conditions." As the final touch to his amygdala workout routine, Ogas developed a meditation technique to use between rounds in order to let his brain cool down: he practiced moving his awareness to his breath while repeating a calming mantra—"nothing weird or chanty," he assured me.

Like the rest of Ogas's training, much of this was pretty radical stuff, far too severe for general consumption. Yet sometimes it's in the extremes that we see the truth most clearly: developing solid cognition under fire isn't about thinking fearlessly but about thinking *alongside* fear. After two months of full-time trivia study, public humiliation, amygdala whispering, and spousal-patience testing, Ogas was finally ready for game show redemption. It was time to play *Grand Slam*.

Lex Luthor Meets Superman

The moment that he first set foot on *Grand Slam*'s disco-dance-floor stage, Ogi Ogas felt a dreadful premonition that the triumphant game show comeback he had imagined was about to flop miserably. Surrounded once again by television cameras and a buzzing audience, Ogas flashed back to the *Millionaire* hot seat, his amygdala triggering a cloudburst of fear. But this time around, Ogas had a few new arrows in his cognitive quiver. As the show's producers prepared to roll, Ogas slipped his eyes shut and began repeating his mantra in his head. His heart rate slowly decelerated. He let the adrenaline vent out of his system. By the time the lights came up and the producers cued him to step forward, Ogas had settled into the posture he would maintain throughout his *Grand Slam* campaign: arms crossed, face blank, eyes straight ahead in a steely stare.

Ogas's adversary in the first round of the tournament was Nancy Christy, an eighth grade English teacher and former *Millionaire* champion. From the opening seconds of their match, it was abundantly clear that Christy hadn't spent nearly enough time bellowing the names of U.S. presidents in department stores; in terms of preparation, Ogas was simply on another planet. As Ogas banged out answers—most of which came straight out of his quick-access Tier One memory bank—Christy struggled to think under the ticking clock. After losing the general trivia round by sixteen seconds, Christy became flustered during the math round, missing a slew of consecutive questions like "Divide the number of maniacs in Natalie Merchant's band by the number of guys who made up *NSYNC."* Meanwhile, Ogas maintained the same unnervingly serene gaze. "He's a machine," Christy told the cameras backstage with an amused shrug. When the dust cleared at the end of the final round, Ogas had demolished Christy by 110 seconds.

His next test wouldn't be so easy. For his quarterfinal match the following afternoon, Ogas drew an opponent whose very name could freeze the blood of lesser contestants in their veins: Brad Rutter. "Everyone there was terrified of him," Ogas recalled. "The man's a monster. It's unsettling." Let's take a moment to review Rutter's resume. Back in 2000, at just twenty-two years old, the lanky and amiable Rutter made his *Jeopardy!* debut with five consecutive wins, which was then the show's limit before a contestant got cycled off. But over the next few years, Rutter cruised through one *Jeopardy!* tournament after another—including 2005's 145-contestant "Ultimate Tournament of Champions," in which he crushed several multichamps like

---

*Answer: 2,000.

Ken Jennings on his way to the $1.5 million grand prize. By the time he met Ogas on *Grand Slam*, Rutter had never lost a game show. Most consider him the greatest contestant in *Jeopardy!* history, and his $3,270,102 in winnings made him the top American game show earner ever.* In short, Brad Rutter was a trivia Superman.

But to Ogas, the most alarming challenge Rutter presented was this: he seemed to perform *better* under stress. "I don't think I'm any smarter than anyone else who's done well on *Jeopardy!*—I just think I handle the pressure better," Rutter told me. Rutter believes the rush of appearing on game shows helps his brain reach new heights. "There's stuff I come up with when I'm actually playing *Jeopardy!* that I wouldn't get if I were just watching it on TV," he said. "The adrenaline focuses me." As an example of this, Rutter points to an "Ultimate Tournament of Champions" match that he calls "the worst game I ever played." When Final Jeopardy arrived in this contest, Rutter sat in a distant third place, with only a sliver of hope left for victory. Alex Trebek delivered the clue: "This Mediterranean island shares its name with President Garfield's nickname for his wife." The question baffled Rutter. "I said, 'Okay, what's a Mediterranean island that sounds like a girl's name?'" he recalled. "I wrote 'Malta' down, but I just knew it was wrong." Suddenly, a vision of a page filled with the names of America's First Ladies leapt into his mind. "I saw the name 'Lucretia,' and with two seconds left I crossed out 'Malta' and scribbled down 'Crete,'" he said. "Lo and behold, the other two got it wrong and I got it right—and I know the adrenaline got me there." Rutter won the game by a single dollar.

Rutter's preternatural focus under fire gave Ogas nightmares;

---

*Rutter has since been surpassed by Ken Jennings, whose winnings total reached $3.6 million after an October 2008 appearance on *Are You Smarter Than a Fifth Grader?*

he considered this bout to be the ultimate test of his cognitive training. As the two stepped forward to face each other, the disparity between them was plain to see: Rutter, the natural-born supercomputer, lazily slid his hands into his pockets while Ogas, the obsessive schemer, folded his arms and projected his diabolical stare. The match began inauspiciously, as Rutter quickly rattled off a spree of correct answers and took the general trivia round by nineteen seconds. A flutter of concern creased Ogas's brow. In the numbers and logic round, though, Ogas was in rare form. When the announcer asked, "What does x equal in this equation: $3x + 10 = 2x + 50$," Ogas shot out the correct answer (forty) almost instantly, while Rutter, a self-admitted math weakling, struggled horribly. Ogas took the round by a massive thirty-nine seconds. In a break backstage, a smirking Ogas told the cameras, "Brad does math like a little girl. There's no chance he's going to take the game away from me." Yet in the third round Rutter stormed back, even as Ogas loosed his craftiest cognitive tricks. "Brad Rutter—he was cold as fucking ice," Ogas told me. "That guy does not get rattled. He was like a goddamn lion prowling after me." Rutter won round three by fourteen seconds.

The decisive final round, then, would present an even battle: Ogas carried in thirty-nine seconds while Rutter kept thirty-three seconds. Of course, in the comic books, this is the part where a resolute-looking Superman summons strength from his inexhaustible wellspring of manly vitality and dashes Lex Luthor's plot for world domination. Yet suddenly, something seemed very different on that *Grand Slam* stage. Superman was *nervous*. As Ogas—the guy who couldn't sit still in the *Millionaire* hot seat a year before—maintained his blank gaze, Rutter fidgeted and rocked on his feet. He grimaced at each missed question. His eyes darted up to his timer. In short order, Ogas built a forty-five-second lead and hung on for dear life. Rutter made a lightning-quick late push, but it was too late. Once Rutter's timer reached zero, Ogas thrust his fist in the air and executed a goofy, elated pivot, screaming "Yes!" just as ferociously as he had bawled "Damn it!" after the final *Millionaire* question. Ogi Ogas, the neurotic grinder, had just become the first person to best the highest-earning game show champion in American history. "All the preparation I did just worked perfectly there," Ogas marveled later. "When I got up there and played, I never felt the stress or the pressure. It was the opposite of *Millionaire*."

After his clash with Rutter, Ogas's semifinal match against the *Twenty-One* virtuoso David Legler was like a stroll on the beach. In both of the first two rounds of their bout, Ogas answered all of his questions correctly—and, adding nerd insult to nerd injury, he delivered gratuitous fist pumps after each victory, hoping to unnerve Legler. It worked. Ogas crushed him by a minute and a half. The scheming neuroscientist had just overcome three trivia titans who boasted a combined total of $6,035,102 in past game show spoils. And now only one man stood between Ogas and sweet redemption: the most famous trivia champion on the planet, Ken Jennings.

Superman II: The Revenge

If God himself elected to spend a leisurely afternoon creating the perfect game show contestant, even he would be hard-pressed to design a better human trivia machine than Ken Jennings. Boyish and unfailingly pleasant, with a giant sponge of a brain, Jennings lacked a single weak subject. Plus, after his seventy-four consecutive *Jeopardy!* victories in 2004, competing before a live audience left Jennings utterly unfazed. He was like a genial Mormon wrecking ball.

In the first two rounds of his *Grand Slam* matchup with Ogas, however, it looked like Jennings had finally smacked into an immovable object. Ogas and Jennings both began at a blazing pace, matching each other step for step: when their first-round firefight ended, Ogas had come out on top by a mere 0.14 seconds. The crowd whooped its approval. The math and logic round likewise ended in a dead heat. For a sublime few moments, it seemed that the scrappy scientist might just defeat the nation's top trivia master and achieve blessed release from his *Millionaire* meltdown.

And then the wheels started to come off. There were two reasons for this. First, to lengthen the final game, the *Grand Slam* producers added an extra round on pop culture. Ogas, he of the zero television sets, got annihilated here; when asked "What star of the Showtime series *Fat Actress* claims to have lost seventy-five pounds using the Jenny Craig program?" (answer: Kirstie Alley), Ogas looked as bewildered as if the announcer had read the question in Romanian. But to Ogas, the real culprit behind his unraveling was something out of his control: scheduling. "They told me they'd be taping the show in the morning, so as part of my training I set up a rigorous sleep cycle," he explained. "Every day, I'd get up at five a.m. and go to bed at ten p.m., so my mental activity would be at its peak when I was expecting them to tape." But when Ogas arrived, he was shocked to learn that tapings would be in the evenings instead. For three nights, Ogas had tried to snap into a new circadian rhythm, forcing himself to stay awake until two a.m. with frigid showers — but he kept waking up at five thirty in the morning. By the time he faced off against Jennings it was nearly midnight, and Ogas felt so exhausted that he considered faking an illness to postpone the match.

With Jennings commandingly ahead going into the final round, their last clash was merely a formality. Jennings pounced with the same intensity as before while Ogas looked on, deflated. As Jennings grinned at his victory and the $100,000 purse he'd be taking home, Ogas issued an irritated sideways glance: his cognitive mission had come up just short once again. "Second place won zero—it was winner take all," he told me with a sigh.

But Ogas admits that even though he failed to capture *Grand Slam*'s top prize, his campaign was a rousing success on one front: he'd transformed himself into an expert thinker under pressure. By delivering embarrassing fact-announcements on the streets of Boston, he shrank his fear of crowds. By training himself to think flexibly when anxious, he learned to ignore irrelevant worries and feelings of fear, focusing instead on the job in front of him. With a little bit of work—okay, a stunning amount of obsessive over-preparation—Ogas swung from helmet fire to impressive cool-headedness. "If you watch my million-dollar question on *Millionaire* versus the final against Ken Jennings on *Grand Slam*, it's night and day," Ogas said. "I was out of my mind on *Millionaire*, but I'm just stone cold against Jennings. All of the training really did work." His newfound poise has also trickled into other pursuits. "The thing that turned out to be most useful was the relaxation stuff," he told me. "Before, if I had to give a lecture in my field, I'd feel nervous and neurotic—my heartbeat would go up, my hands would get clammy. But now I'm completely comfortable with it. When I first start feeling pressure, I automatically go into the breathing thing. I can get my mind back into that calm state." (We'll explore this useful breathing tip more in Chapter Seven.)

"Let me just get this out: it still burns," Ogas reflected, gritting his teeth. "Not as much as *Millionaire*, because it was out of my hands, but it still burns." Once again, he finds himself praying for another shot at game show redemption. But nowadays, he isn't seething about imploding under pressure; he just wants to claim what was rightfully his. "Look, Ken was the best player there besides me," Ogas told me, "but I'm pretty sure I would have beat him if I'd played him in the morning.

"It should have been me," he murmured. "God*damn*."

CHAPTER FIVE

Before the Madding Crowd:
Decoding the Mysteries
of Performance Anxiety
and Stage Fright

T he human brain is a wonderful thing," the comic actor George Jessel quipped many years ago. "It starts working the moment you are born, and never stops until you stand up to speak in public." It's a good line—as befits a man once known as the "Toastmaster General of the United States"—but Jessel's wisecrack isn't quite accurate. We might *feel* like our mental circuitry ceases to function when the ravenous eyes of an audience lock onto us, but at least one section of the neural mainframe is working at warp speed: the amygdala.

As a species, we Homo sapiens play host to an astoundingly vigorous fear of appearing or performing before groups of other humans, be it second grade show-and-tell or a concerto performance at Carnegie Hall. A wide variety of studies have crowned fear of public speaking—or glossophobia, for sticklers—as our king of all phobias; according to a 2001 Gallup poll, more than 40 percent of Americans confess to a dread of appearing before spectators. (In some surveys, fear of public speaking even outranks fear of death, a fact that inspired Jerry Seinfeld's famous observation that at a funeral, this means the average person would rather be in the casket than giving the eulogy.) Our terror over audiences has a solid evolutionary foundation, of course: in the natural world, whenever you notice several sets of eyes scrutinizing you intently, they're probably estimating how tasty you'll be as the evening's main course. Present-day crowds generally carry little, if any, carnivorous intent into their spectating, yet still, whenever the spotlight beams our way, few among us can evade the pallid horror of performance anxiety.

But what if we could change this and stack the odds in our favor a bit? Let's run a quick thought experiment. Imagine that you have somehow been persuaded to publicly perform a soliloquy from *Hamlet*, but you've also been granted the magical ability to design the ideal audience to minimize your stage fright—a crowd custom-built to soothe your amygdala and spur you on to peak performance. What might we expect this audience to look like?

To begin with, we'd no doubt want to stock your perfect crowd with people who make you feel comfortable: your spouse, friends, family, and other well-wishers. Surely, being surrounded by a friendly audience will ease your performance anxiety, right? Actually, no. As the psychologists Roy Baumeister and Jennifer Butler demonstrated in one 1998 study, performing in front of friends brings unforeseen pitfalls. For their experiment, Baumeister and Butler sat participants in front of a one-way mirror and assigned them a challenging arithmetic task: beginning at the number 1,470, rapidly count backwards in chunks of 13 (i.e., 1,470... 1,457...1,444...1,431). They told half of their forty subjects that a stranger was watching on the other side of the mirror; the other half learned that a friend sat behind the glass. Surprisingly, the participants who thought they were performing in front of their pals produced 30 percent *fewer* correct responses than those who thought strangers were watching. Baumeister chalks this discrepancy up to the "friend" group's anxiety about disappointing their chums, which made them more cautious and tentative. "When your friends and family are watching, you feel a lot more self-conscious," Baumeister told me. "You're more aware of being watched, and that attention to self disrupts your performance."

All right, so a crowd packed with loved ones is out. But if you need to perform in front of a bunch of strangers, certainly they should express a little support—maybe wave signs with slogans like YOU CAN DO IT, READER! and READER IS #1! I mean, it would be bad if these people were *totally* ambivalent, wouldn't it? Wrong again. Although Baumeister concedes that supportive audiences can help boost your motivation to succeed, which is helpful in strictly effort-based tasks like marathon running, they're often detrimental in more skill-oriented pursuits because they tend to swing your performance strategy in an unhelpful direction. As Baumeister wrote in one paper, "Audience support magnifies performance pressure and induces performers to avoid failure rather than seek success." (Expressing high expectations often had the same subconscious anxiety-provoking effect.) One Baumeister study even found that complimenting a performer on something completely unrelated to the task at hand—the example they tested was "That's a nice shirt you're wearing"—negatively affected his or her performance on a skill test. "Again, praise makes people self-conscious, and that hurts their ability to perform," Baumeister explained.

So your amygdala-friendly crowd needs to be laden with neutral, unsupportive strangers, perhaps toting signs saying WE HAVE NO OPINION ON YOUR SHIRT. Okay. But can't the members of this audience at least be *human?* Well, according to research by the UC Santa Barbara psychologist Jim Blascovich, probably not. In the early 1990s, Blascovich hooked female volunteers up to cardiovascular measurement equipment in their own homes, to gauge their stress level as they carried out a taxing arithmetic task under three very different audience conditions. First, they established a baseline stress score while the women tried the task alone. Next, the subjects attempted the math exercise in front of their closest female friends. "The best friend just threw the stress level off the chart," Blascovich recalled. "They felt much more threatened." Finally, Blascovich had the women perform the math task in front of a far less analytical spectator: their pet dog. "What we found was that the least stressful, threatening situation of the three *by far* was if you had the dog there," Blascovich said. "I couldn't believe it. I'm not a pet person." One of Blascovich's volunteers couldn't believe it either. A few weeks after the study concluded, she called Blascovich and asked him to repeat the test, this time with her beloved husband in the room. He obliged. "The spouse made the stress level go a whole lot higher than the dog," Blascovich told me. "Our explanation was that dogs are unconditionally loving and nonjudgmental—which made us think this would never work with cats. But it turns out it works with cats, too."

Let's summarize our findings, then, shall we? In order to minimize stage fright for your *Hamlet* speech, you'll be delivering it before an optimized audience of unfamiliar, ambivalent house pets, many of whom will be displaying banners declaring their modest expectations for your performance. While this sounds like a recipe for a truly memorable afternoon, it's safe to say that this isn't a crowd we're likely to encounter anytime soon. Instead, we're stuck with the tough crowd—the kind of amygdala-rattling audience that blanks our brains, sends us into clammy sweats, and releases butterflies by the dozen into our gastric regions. Virtually by design, crowds are destined to make most of us want to flee the vicinity in terror.

So it was, once upon a time, with the cellist Zoe Keating. These days, Keating works at the vanguard of experimental classical music, and her innovative playing style has won her a wide, enthusiastic following; her music has been featured everywhere from National Public Radio to film scores to European ballets. For her inventive approach to composition, Keating is often described as a "one-woman string quartet." Using only a laptop computer, an electronic looping pedal, and her cello, Keating builds mesmerizing orchestral suites layer by fluid layer, the individual voices flowing in and out with a twitch of her bow or a tap of her foot. One winter evening not long ago, I watched Keating hold a large crowd rapt as she performed inside a cavernous ballroom. Under the kaleidoscopic stage lights, she cut a striking figure. Ghost pale in an elegant Victorian costume, a thatch of red dreadlocks pluming up from the top of her head, Keating seemed entirely oblivious to the hundreds of eyes watching her. She played as though she were in the midst of a dream, eyes closed, swaying languidly with her cello, utterly immersed in her performance. "Usually when I perform I get this tunnel vision effect where I have no idea how much time has passed," Keating told me later. "Whether I've been playing for a minute or an hour, I couldn't tell you."

Keating performed with such absolute composure that night, in fact, that it's difficult to believe she once suffered from the most debilitating stage fright imaginable—a crippling case of anxiety that made her quit performing music entirely. For Keating, the breaking point came during a recital when she was a seventeen-year-old cello prodigy in upstate New York, already accepted (full ride included) to several of the nation's most prestigious conservatories. "It was a big deal concert—you know, five hundred people, that kind of thing—where I was a soloist," recalled Keating, who is now in her late thirties. "I got up there on stage and I was so nervous that I dropped my bow. And suddenly *I could not remember how to play the cello.*" The recital was an unmitigated catastrophe. Stewing in her teenage dread, Keating believed her intense performance anxiety meant an untimely and irrevocable end to her music career. "I actually didn't even know you *could* get over

169

stage fright," she said. "I just thought, if you can't handle it, you shouldn't be on stage."

And yet, many years and many painful false starts later, here Keating is, cello in hand, perfectly poised, alone before the murmuring throng and reveling in the experience. The solution to her stage fright, it turned out, was far simpler than she had imagined.

Fright Club

In the mid-afternoon of February 3, 2002, the New England Patriots quarterback Tom Brady sat in the locker room of New Orleans's Louisiana Superdome and awaited the call to storm out for the start of Super Bowl XXXVI, now only hours away. At just twenty-four years old, Brady was one of the youngest quarterbacks ever to play in a Super Bowl; indeed, as of five months before, he'd barely even set foot on the field in a professional game. Now he was about to play before a television audience of hundreds of millions, and against the massively favored St. Louis Rams, no less—an offensive juggernaut nicknamed the "Greatest Show on Turf." If ever there was a time for a blue-ribbon-winning fit of nerves, this was it. So, as the pregame show began and the nation geared up to watch him in the year's biggest sports spectacle, what did Brady do? He took a refreshing nap. Let me just repeat that: a *nap*. And this astounding lack of stage fright soon carried over into the game itself. That evening, Brady led the Patriots with almost supernatural poise, coolly marching his offense down the field for a game-winning field goal as time ran out, and taking home the game's Most Valuable Player trophy. It's the kind of fearless tale that makes a mortal man into a sports god. It's also very, *very* unusual.

From the amateur ranks on up to the planet's most elite performers, stage fright is more common than most of us realize. Not even the legends are immune to its thorns. As a counterpoint to Brady, consider the basketball icon Bill Russell. No athlete has ever dominated a sport as thoroughly as Russell dominated basketball in the 1950s and '60s. Over a span of fifteen seasons, Russell won two NCAA championships with the University of San Francisco, one Olympic gold medal, and a staggering *eleven* NBA titles with the Boston Celtics. Russell was, to sum up, a man with nothing left to prove. And yet throughout his decorated career, he seldom managed to duck his famous pregame nerves. "Night after night, sixty or seventy times a season, [Russell] was in the head before the game tossing the remains of his lunch," wrote John Taylor in his book *The Rivalry: Bill Russell, Wilt Chamberlain, and the Golden Age of Basketball.* "In fact, he did it with such regularity that to his teammates it became a ritualistic sign of good luck. If Russell wasn't in there puking, they got worried."

Perusing the biographies of great performers yields countless stories like this, in which artists and athletes endure punishingly intense nerves for their craft. The thread begins all the way back in ancient Rome, when the illustrious orator Cicero admitted of his celebrated performances, "I turn pale at the outset of a speech and quake in every limb and in all my soul." Ella Fitzgerald suffered such strong preconcert anxiety that she had to station herself in a specific spot beside the stage just before going on, where fellow musicians say she carried out a strange and elaborate set of ritualistic movements to prepare her to vault in front of the crowd. The legendary tenor Luciano Pavarotti, who famously needed to find a bent nail somewhere on the set and place it in his pocket before every performance, likewise endured a powerful lifelong case of the butterflies. "What I go through in the last ten minutes before a performance I wouldn't wish on my worst enemy," Pavarotti claimed. In some cases, nerves have driven celebrated performers off the stage entirely; the singer Barbra Streisand dodged live performance for nearly thirty years after a traumatic 1967 concert in Central Park, in which she blanked on some of her lyrics.

But probably the most storied case of stage fright on record belonged to a man widely considered the greatest actor of his era: Laurence Olivier. For Olivier, the fear first struck when he was a somewhat vain fourteen-year-old theater prodigy at the All Saints Choir School in London. As he relates it in his autobiography, *Confessions of an Actor*, Olivier was just about to sing his swan song in a Trinity Sunday choir performance when he "yielded to the wicked inclination to make a boastful gesture for the benefit of the rest of the choir." Attempting to appear dashing and possessed of unfathomable depths of soul, Olivier tilted his head "at an aloof angle toward some imagined distant clouds" and began to sing. After only a few bars, though, intense guilt over his narcissism weighed down on him. Soon, he writes, "my voice faltered; the breath left my being and could not be retrieved; my throat closed up and I was forced to stop." Olivier gutted his way through the rest of the song, but the incident planted a seed of self-doubt that would linger throughout his remarkable career.

Off and on for the next fifty years, stage fright dogged Olivier. "He is always waiting outside the door, any door, waiting to get you," he once declared. In 1964, for example, when Olivier was starring in a production of Ibsen's *The Master Builder*, a sudden dread took hold of him on opening night, convincing him that he would forget his lines.* "With each succeeding minute it became less possible to resist this horror," he later recalled. "My cue came, and on I went to the stage, where I knew with grim certainty I could not be capable of remaining more than a few minutes." For

---

*Side note: one U.K. study found that on the opening night of a play, the actors' average stress levels are roughly similar to those of a car accident victim.

a terrifying instant, Olivier once again felt his throat constrict. The audience spun in circles before him. But then, he spotted the legendary Noel Coward sitting in the front row. Instead of making him more anxious, this triggered a feeling of even greater intensity: his ferocious competitive pride, which vaulted him right back into the action. As always, he scarcely missed a beat; no one noticed anything had been amiss.

To deal with his fierce performance anxiety, Olivier often enlisted the help of his fellow actors. When he performed the title role in *Othello*, to rapturous reviews, Olivier convinced Frank Finlay (playing Iago) to stand in the wings whenever he was delivering a soliloquy, so he wouldn't feel alone with the crowd. On another occasion, Olivier asked his colleagues not to look him in the eye while on stage, because it made him uneasy. (They looked at the side of his face instead.) The theater director Peter Hall once watched in amazement as a quivering Olivier prepared for an entrance a few feet off stage—and in the clutches of a stage manager. "Just before his cue Larry said, 'Right, push, push, push,'" Hall told the Olivier biographer Terry Coleman, "and the stage manager pushed Larry on stage, and one saw Larry walk on stage and become Olivier."

Yet despite the pain his stage fright caused him, Olivier dealt with the strife admirably, even inspiringly: he never let anxiety deter him from his calling. "There was no other treatment than the well-worn practice of wearing *it*—the terror—out, and it was in that determined spirit that I got on with the job," he wrote in his autobiography. In 1970, after weathering an especially intense bout of nerves, Olivier began making peace with his stage fright; finally, at age sixty-three, he was surprised to find that his fear was fading away. Olivier's extraordinary saga—alongside the stories of those like Russell and Pavarotti—points us to an important truth that those with performance anxiety often miss: even when we're experiencing an insistent case of nerves, we can still perform at an incredibly high level. Stage fright is unpleasant, sure, but it's a workable predicament.

Back in her tortured teenage years in upstate New York, though, Zoe Keating believed that just the opposite was true. Stage nerves first hit Keating when she was fifteen years old—right when music becomes a brutally competitive dogfight for serious young orchestra musicians—and from that point on, each public performance felt like a battle for psychological survival. "At auditions, I would freeze to the point of not being able to play the cello," she recalled. "Then there's the classic thing where your hands get so wet that you can't hold on to your bow—it's like they're suddenly covered in soap. Or I would shake. And the whole performance was about trying to keep it together and not lose my shit." Keating's talent usually managed to shine through during these grisly ordeals, yet she knew her nerves were obscuring her true ability. "If you're so busy trying to keep yourself from passing out in terror, that doesn't allow you to be very musical," she said. As Keating worked her way up the ladder, the auditions grew more and more frequent. "They're always increasing in intensity," she explained. "By the time you're sixteen years old, you're auditioning for concerto competitions and regional orchestras made up of the best musicians in a five-state area. Or you're auditioning to solo in front of an orchestra, or for music camps. Plus, if you're in an orchestra, every few months you have section auditions to decide the order of the seating. So you've constantly got someone sitting in front of you, staring at you, judging you." These auditions were so traumatic for Keating, she says, that she can't remember many of them at all: "They're blocked out."

Like many developing musicians with performance nerves, Keating received nothing but unhelpful advice when she sought counsel for her stage fright. "When I mentioned it to my cello teacher, she said that if I was having any anxiety, it meant I hadn't practiced enough," she recalled. Keating took this opinion as gospel, yet there was one problem with it: ever since she first took up the instrument at age eight, Keating had lived and breathed cello. "I did nothing *but* practice," she told me, frustration still evident in her voice. Cello was her entire world. Youth orchestras and music camps made up the bulk of her recreational schedule. Several times a week, Keating's mother drove her thirty miles to Rochester, where she attended the preparatory program at the prestigious Eastman School of Music. Alone in her room, Keating was as polished a teenage cellist as they come. "When I was on my own, I could play these difficult pieces backwards and forwards perfectly," she said. "So there was this incredible frustration where I knew I was really good, but once I actually got in front of people I couldn't *prove* I was good." Even as Keating ratcheted up her practice time, the auditions and concerts got no easier.

Paul Salmon, a clinical psychologist at the University of Louisville who specializes in treating stage fright, says unproductive anxiety tips like the one Keating received are distressingly common in the musical world. "We know a lot about the physiology of stress and anxiety nowadays, but it hasn't made its way into musical circles," Salmon told me. "Here at the university, we have a music school where I've done some consultation, and people there have very strange ideas about how to deal with stage fright—or no idea at all, which is even worse. Sometimes, they don't have any thoughts about it other than 'practice more.'" Because the issue anxious performers face is cognitive and emotional, not a matter of mechanical technique, Salmon says that adding routine practice time won't fix much—a view that echoes the performance expert Anders Ericsson, whom we met in Chapter Three. For already hyper-rehearsed musicians like Keating, running through a Bach cello suite ten more times at home won't shift their feelings about performing in front of others. It's like telling someone who's afraid of flying to deal with it by practicing sitting in a chair; you're not exactly getting at the heart of the dilemma.

Even compared to other harrowing tasks like stage acting and public speaking, Salmon believes classical musicians face the toughest, most anxiety-provoking challenge of all. In a world where audiences have grown accustomed to the crisp, note-perfect performances they hear in digital recordings, classical musicians feel ever-greater pressure to play flawlessly. "In a speech, no one knows what you're planning to say," Salmon explained. "But with music, you're less protected. You're out there playing from memory a lot of the time, and you're supposed to maintain complete fidelity. You're always reproducing something. When I go to concerts, I often see people following along with a score; they're not listening to the music, they're just checking off notes." In many ways, auditions are even harsher on a performer's amygdala: audition boards go to great lengths to screen out anything related to a musician's personality, focusing only on technique and playing ability. Joshua Aronson's research on test anxiety, you'll recall, found that reminding students that they're multitalented, whole people helps reduce their performance worries—yet the audition process sends an opposite, and hence quite fear-provoking, message. "We care about your musical skills," it says, "not about you."

It was this unwavering demand for perfection that Keating felt crushing down on her whenever she performed. Over the course of two years, even as she won more recognition as a cello prodigy, her stage fright seeped into every corner of her life. "It got to be this sort of pervasive sense of doom," Keating told me. "I'd be going through my normal day, and the only thing I could think was *Oh my God, I have an audition in seven days.* It created this dark shadow behind everything." She had once fantasized about becoming the conductor of a major symphony; now she wanted only to escape. By the time Keating arrived at that fateful, anxiety-racked moment of soloing in front of a large recital crowd, her dreams were already teetering on a knife's edge. Years later, sifting through some old cassette tapes, Keating found a recording of the performance, and the audio evidence confirmed her teenage mortification: it was, indeed, really bad. Her shaking hands sent the bow clattering to the ground. She had to stop and start again. "It's this professional scenario where I'm supposed to be this great young musician, and I sound like I'm maybe ten years old, sawing at the cello like an amateur," Keating said. "When I finally stop, you can hear a pin drop in the silence." She left the stage beaten and humiliated.

No one said anything to Keating about her meltdown that day. And, just as quietly, she decided she'd had enough of her stage fright. Keating had three conservatory scholarships waiting for her, and she turned them all down. In the symphony world, opportunities for performers who haven't attended a topflight conservatory program are virtually nonexistent; Keating was shutting the door on her dreams. "I couldn't solve my stage fright, so I just assumed that was because I couldn't handle performing," she explained. It would take her many years to figure out just how wrong she had been.

Cracked Shot

In August 2008, officials at the Beijing Olympics announced that doping tests had fingered a multi-medal winner at the Games, Kim Jong-su of North Korea, for using a banned performance-enhancing drug. With a hailstorm of derision about besmirching the sanctity of the Games, the International Olympic Committee stripped Kim of his two Beijing medals (a silver and a bronze) and exiled him from the Olympic Village—another cheater dealt with swiftly. But those who read beyond the headlines found something peculiar about Kim's case. Kim hadn't tested positive for steroids or endurance-boosting hormones. He wasn't a weightlifter, a cyclist, or a sprinter. Kim was a pistol shooter, hardly the type of athlete who needs extra muscular bulk. The drug he'd been busted for taking was the beta-blocker propranolol, an anti-hypertension drug we learned about earlier (taken right after an accident, it can inhibit the formation of traumatic memories). Because propranolol blocks adrenaline, it also counters some of the physical effects of anxiety, like shaky hands, rapid pulse, and sweaty palms. Essentially, Kim was guilty of unfairly rigging his stage fright.

In a fine-tuned, precise sport like shooting, even a mild case of performance nerves can mean the difference between a gold medal and a broken heart. Take the curious tale of Matt Emmons, another shooter who competed in Beijing. A goateed and friendly twenty-nine-year-old from rural New Jersey, Emmons is, by a fairly wide margin, the best rifle shot on the planet. Shooting often gets grouped with fringe athletic pursuits we barely consider "sports," like table tennis and curling, yet the challenge shooters like Emmons face is immense: from fifty meters away, half the length of a football field, they aim to hit a bull's-eye smaller than a dime. Lying prone with his .22-caliber rifle in competition, Emmons is one of only a handful of shooters in history who has nailed this bull's-eye in every single one of his sixty preliminary-round shots, for a perfect score of 600. The man can shoot.

The goal of a marksman, unlike virtually any other breed of athlete, is to remain as still as humanly possible as he locks onto the distant bull's-eye. For the University of Maryland sport psychologist Bradley Hatfield, who has conducted several studies on the brains of elite shooters, this represents an extraordinary feat. "It's an absolute study in concentration," Hatfield told me. "They don't move. These guys are like machines." Naturally, this stillness requires tremendous self-control; topflight shooters dread catching colds, because the slight uptick in heart rate could throw off their razor-sharp physiological equilibrium. (Not that this hindered Emmons: at the 2004 Olympics in Athens, he endured a cold throughout the fifty-meter prone competition but took the gold anyway.)

At the 2008 Games, Emmons arrived in Beijing with something to prove. Four years earlier in Athens, he had built a massive lead going into his final shot of the three-position rifle event. To earn his second gold of the Games, he just needed to hit *anywhere* on the target. A portrait of focus in his science-fiction-esque shooting garb (all marksmen look something like *Star Wars* extras), Emmons drew down and fired what he believed was a solid shot. But, strangely, his score failed to come up. Emmons turned in confusion to the judges and said, "I shot," producing the spent cartridge as proof. And then, a sickening realization sank in: Emmons had fired at the target in lane 3. He was in lane 2. "That'll make a hell of a story," he muttered. He scored a zero for the shot, and, in an instant, fell from first place to eighth. Emmons chalked the mistake up to simple bad luck. "Looking at the target number is just something you never do," he told me. "It was a freak accident."

After snaring a respectable silver medal in Beijing's fifty-meter prone rifle event, Emmons once again breezed through the preliminary rounds of the three-position contest in 2008. The first nine shots of the ten-shot final "went like clockwork," Emmons recalls. Once again, he carried a titanic lead into the final shot of the competition. But picking up his .22 for his decisive try, Emmons looked tense. His breathing seemed labored. "I'd thought about that final probably every day for three months," he told me. Emmons double- and triple-checked his lane number ("I wasn't going to make that mistake twice," he said) and began to draw down on the target. "Before the shot, I was nervous," he recalled, "but as I looked through the sights and was relaxing down into the bull's-eye, I remember vividly that I got this sense of confidence, like it's going to be okay. It's going to be okay. And when I touched the trigger, the gun just went bang."

Emmons's quivering hands made him fire too early, the bullet landing well above the center of the target. "I didn't think my finger was shaking," he said. "I guess it was." He scored a 4.4. The aftermath of the shot is tough to watch: Emmons drops his right hand and gazes at the ground; his wife, in the stands, goes wide-eyed with shock, struggling not to cry; the crowd erupts with joy as it realizes the Chinese competitor, Qiu Jian, has made an improbable leap into first place. Emmons fell to fourth. "I just remember thinking, *Damn it, not again,*" he said. Emmons congratulated the winner with a smile and a hug. In interviews, he shrugged affably and said it just wasn't meant to be. But the defeat tore at him. Today, Emmons admits that nerves played a role in both of his last-shot mishaps, but he maintains that he's mostly suffered from bad luck. "It drives me crazy when people say I choked," Emmons said. "Honest to God, I wish I could have walked out and told all of those reporters that I was crapping my pants and I couldn't handle it, but that's just not the truth. Yes, the nervousness was a little bit higher. But everything was going according to plan and then suddenly *POOF!* The gun went off."*

It's one of stage fright's harshest ironies: the first things we lose when anxious are precisely the things elite performers need

---

*Emmons is correct in saying he didn't choke, by the way, but for more complex reasons than he probably thinks. Choking on skill-based tasks like sports, which we'll explore in depth in the next chapter, is different from the simple performance anxiety Emmons felt; performance anxiety often *leads* to choking, but choking is also a discrete physiological process that goes far beyond anxiety. In other words, feeling anxious and having your hand shake—which is what foiled Emmons—isn't technically the same thing as choking. Again, more on this soon.

most—especially fine motor control. When the brain's fear system revs up, adrenaline directs oxygen and glucose into the large muscle groups that enable us to fight or flee, disrupting our more intricate motor skills. Hands become shaky and clammy with lack of blood flow. Muscles in the throat and vocal cords tighten, making the voice waver and swing ever higher in pitch. (In extreme cases, this produces the piercing scream that works so well for women in horror movies but that results in immediate and irretrievable loss of dignity in startled men.) Heart rate and breathing quicken, providing the body with more oxygen but dealing another blow to fine motor control. Digestion and saliva production shut down, bringing on butterflies and cotton mouth.

For a shooter like Emmons, an actor like Olivier, or a musician like Keating, these stage fright symptoms can directly interfere with the ability to perform. If anxiety makes your hands quiver, it's tough to shoot straight or play a difficult cello suite; if it makes your voice sound like Barney Fife's (which, fortunately, never happened to Olivier), you can hardly deliver the grave lines of a character like Othello. As we saw in Yerkes and Dodson's research last chapter, *some* anxiety is necessary to perform your best, yet increasingly complex tasks are decreasingly forgiving to nerves. "A sort of minimal level of anxiety will promote performance," explained Michelle Craske, the UCLA anxiety expert. "It can be very powerful. But the harder and more intricate the task that you're dealing with, the lower that optimal level of anxiety should be." Matt Emmons's Beijing shooting final called for the most delicate motor skills the human body can produce, which was why even a split second of nervous shakiness could send him to gut-wrenching defeat.

Hence the beta-blockers. When the North Korean pistol shooter Kim Jong-su took propranolol, he was trying to evade the same lethal quivering hand that struck Emmons. Because beta-blockers thwart the physical effects of adrenaline, they are the perfect drug for high-pressure shooting; one Swedish research team discovered that they improve performance among marksmen by an average of 13 percent. But beta-blockers are a far more common presence in the classical music world, where their use is both widespread and controversial. In 1987, a survey of more than two thousand professional classical performers revealed that 27 percent of them used a beta-blocker to counter their preconcert nerves. As the competition for a shrinking number of orchestra positions has grown fiercer, use of the drugs has become even more pervasive. Though no current survey data exists, many classical musicians call beta-blockers "ubiquitous." The complaints about their impact on musical performance are legion: some say beta-blockers produce mechanical, emotionless performances, while others deride them as a "crutch." Even more troubling, however, is the reality that so many performers are taking a potent, nervous-system-altering pharmaceutical without consulting their physicians; the FDA has never approved beta-blockers for treating performance anxiety, and most of the musicians taking them do so without a prescription.

Luckily or not, the struggling young Zoe Keating never heard about beta-blockers in her upstate New York cocoon. So over the years, she found another path to deal with her stage fright, a path that involved no drugs—just pure, old-fashioned exposure.

Bach at the BART

Keating might have jumped off the professional classical performance tracks as a teen, but she didn't quit music entirely; she just switched trains. In her undergraduate years at tiny, liberal-arts-oriented Sarah Lawrence College, Keating studied experimental electronic music composition and improvisation, musical disciplines that were as far from classical cello as she could find. Toying around with synthesizers was far less nerve-rattling for her than trying to turn out note-perfect sonatas. "I'd start playing randomly, and I discovered pretty quickly that if the music wasn't written down, I didn't have as much anxiety—I think because there's no expectation of how it's supposed to sound," Keating recalled. Encouraged by the change of pace, she tried a blind audition (where, crucially, no one could see her) for a regional youth symphony and got in. The experience of playing relatively anonymously in the midst of a vast orchestra thrilled her. "It's like you're part of this organic musical whole," she said. "I loved it. I thought I'd finally cracked my stage fright."

After college, Keating moved to the San Francisco Bay Area, where, like so many newly minted liberal arts graduates, she found herself mired in a succession of frustrating dead-end jobs. "So I thought, *This is ridiculous: I have all of this musical skill and the working world sucks—I'll go to grad school for the cello*," she said. "The San Francisco Conservatory of Music has a very good cello program, so I decided to quit my restaurant job and practice for six months, thinking I could get up to speed." By the time her conservatory audition date arrived, Keating brimmed with confidence. And then she walked into the audition room. As Keating set up her cello, the teachers began asking her questions. "Who are you studying with here?" they asked. "No one," Keating admitted. They asked what school she had attended, to which Keating replied, "Sarah Lawrence." "That's not a music school," one teacher pointed out ominously. "My self-confidence was being chipped away, chipped away," Keating told me. "And I felt doubt again." When she finally began to play, she discovered to her horror that she was trembling so badly that she couldn't control her bow. She had to stop and restart the piece. Just as she began to break through the wall, the teachers interrupted her playing. "Well, Zoe, we think you're not quite ready," one of them said. Keating was humiliated. "That felt like the nail in the coffin," she told me.

Dejected, jobless, and without a plan for the future, Keating now faced a different dilemma: she was broke. To make her share of the rent, she needed a quick $250. Out of desperation, she resolved to try to earn some cash by playing her cello during rush hour at the busy Embarcadero and Powell Street Bay Area Rapid Transit (or BART) stations. To say the least, this prospect went over poorly with her newly reignited performance anxiety, yet onward she marched. At first, mortified at having so many potential spectators, Keating stuck to the Bach cello suites she knew best. "I was totally convinced that every single person was analyzing my technique." She laughed. "I'd play the same few pieces over and over because I knew my bow placement was okay, or I was using the correct fingerings." To fall under so many uncaring eyes was horrifying at first. What if they could tell how nervous she was? What if she made a mistake?

But then something unexpected began to happen: commuters *thanked* her for being there. "People would come up and say, 'My mind was so fixated on what a terrible day I was going to have, and hearing you play made my day better,'" Keating recalled. "Even if I'd gotten the technique wrong, people would hand me a five-dollar bill and say, 'That was fantastic!' That was the first sense I ever got that musicians might have a role in enriching the world." Performing wasn't just a pain-endurance exercise, or a battle for perfection; it could actually be rewarding. (The fact that she was pulling in $50 an hour in donations helped amplify this positive feeling.) Soon, she started stretching her wings a bit. "After playing the same thing four times in a row, I'd decide to flip the page to a Bach suite I didn't know so well, and I'd see if I could make it through to the end, just sort of enjoying that I'd never played it before," Keating said. "In other words, I allowed myself to play the music without worrying about all the little things—'Is your shoulder too high? Is your vibrato correct?' And it was *fun*." Over time, as she became more immersed in her musical exploration, her stage fright at these BART station recitals transformed. She still felt nervous, but suddenly it seemed much more manageable—helpful, even. "This is the interesting part," Keating told me. "I can actually *remember* playing in the subway."

Busking in the BART station, Keating was accomplishing a few things, psychologically speaking. Based on what we've already learned, we know that by exposing herself to her fear without running away, Keating was letting her brain slowly habituate to the idea of performing for an audience. Over the hours, as the realization dawned in her unconscious mind that these commuters weren't going to descend on her like starving jackals, her prefrontal cortex taught itself to soothe the amygdala's reaction to the crowd. This principle also explains why programs like Toastmasters are so effective at treating public speaking anxiety: by delivering speech after speech, regularly welcoming their fear, participants learn that standing behind that lectern isn't really a threat to life and limb; it's something they can handle.

But neuroscience aside, Keating was also coming to an important *conscious* insight: her listeners couldn't see through her like she'd thought they could. "These people were anxious to get to work on time," she said. "They didn't care that I had my pinky waving in the air the wrong way." No one really saw her nervousness. If people stopped to listen, that meant they were enjoying the music, not judging her. Keating had finally broken through one of the most pervasive misconceptions underlying performance anxiety, the "illusion of transparency" bias. Put simply, we tend to believe that our internal emotional states are more obvious to others than they truly are. Ask a subject in the psych lab to tell a lie, and she'll believe she's shown more clues about the fib than she actually has; ask her to drink a nasty beverage, and she'll mistakenly think her disgust is fully evident. And, of course, ask an anxious actor like Laurence Olivier to perform before a crowd, and he'll worry that his nerves are practically doing their own tap dance routine out there—even though Olivier always showed dazzling self-command, from the audience's perspective. "If your heart is pounding, you think everyone can see it," said Paul Salmon, the stage fright expert. "When you're physiologically activated, your senses become more highly attuned and you magnify the impact of these things. It's like you're under the microscope." The illusion of transparency bias is powerful, but it's also easy to correct. In one 2003 study, for example, the psychologists Kenneth Savitsky and Thomas Gilovich found that simply explaining to subjects that their stage fright isn't obvious to others made them perform better—and with less anxiety—than two control groups when they gave a speech later on.

Almost immediately, Keating's musical world began to open up. Having reached a better balance with her performance nerves, she felt free to explore. Soon enough, she answered a Musicians Wanted classified ad that read, "Dark wave bass player seeks melancholy strings." She joined bands with gothic monikers like Narcissus and Alfred, and, over time, played her cello in support of progressively bigger musical groups. As she built confidence, she began experimenting with solo compositions that blended her classical training with the electronic music she had studied in college, tapping and tweaking her cello to produce a dynamic range of sounds that she could loop and layer. Today, she makes her living solely through her music, jetting around the world to perform; her first album, *One Cello x 16: Natoma*, has been a steady seller on iTunes since its 2005 release, while her second, 2010's *Into the Trees*, debuted at number seven on the Billboard classical chart. When I spoke to Keating, she was elated over having won a $50,000 Creative Capital Foundation grant earlier that morning, and she still buzzed about a recent stint performing in Valencia, Spain, with a ballet company that had choreographed a program around her music.

Yet even as Keating gained success and grew adept at performing in front of crowds, her stage fright never really left. What did change, however, was her *interpretation* of that fear. For years, as Keating's domineering stage fright twisted her into knots, she could see her performance anxiety only as destructive and incapacitating. As Salmon points out in his book *Notes from the Green Room: Coping with Stress and Anxiety in Musical Performance*, cowritten with Robert G. Meyer, sufferers of performance nerves often catastrophize their fear symptoms (for example, an anxious pianist might become convinced that his rapid pulse augurs an onstage heart attack), and the fact that these predictions virtually never come true doesn't do much to reduce their fear. "Many chronically anxious individuals believe that the more time that elapses without a catastrophe, the more likely one is to occur in the immediate future," they write. Psychologists call this the *debilitative* view of performance anxiety, and it can be a tough perspective for a chronically worried performer to break out of.

Keating, on the other hand, learned that her anxiety could be *facilitative*—it could help her perform *better*. "I still get nervous beforehand," she wrote recently of her concerts, "but it's the kind of healthy nervousness that you'd have to be a stone not to experience. It keeps me on my toes and makes me excited to perform." Indeed, because we need a certain amount of anxiety to perform well anyway, learning to overwrite needlessly pessimistic beliefs about our nerves with the more constructive truth can make a huge difference in how we experience that anxiety—both on and off stage. Psychologists call this process of changing the context of our experience "reappraisal," and it's a fantastic practical tool for transforming our relationship with fear. According to the

NYU neuroscientist and reappraisal expert Elizabeth Phelps, when we take one of our knee-jerk negative views about an anxiety-provoking situation ("Nerves are going to make me blow this presentation") and consciously shift ourselves toward a more positive and realistic perspective ("I'm prepared for this, and fear can help me succeed"), we aid ourselves more than we realize. "Verbal instruction is a uniquely human way to control a fear," Phelps told me. "Learning to reinterpret a situation in a more healthy way actually uses some of the same mental circuitry as fear extinction." Thus, when Salmon is counseling a performer with stage fright, he speaks of pre-performance nerves not in dire terms but in a more affirmative way. "Instead of talking about 'anxiety,' I'll relabel it as 'energy,'" he explained. "Your system is keyed up, your sympathetic nervous system is kicking into gear, and that's fundamentally a good thing. It can give a life and vitality to a performance. Anxiety isn't something to be afraid of."

The move from a debilitative view of performance anxiety to a facilitative one is more than mere sleight of hand. Several studies have shown that a major difference between novice and accomplished performers isn't how much fear they have but how they *frame* that fear. For instance, one 1994 study by a trio of sport psychologists surveyed more than two hundred competitive swimmers and found that there was no discrepancy at all between the elite athletes and non-elite athletes in the intensity of their pre-race anxiety; the elite swimmers, however, were far more likely to describe their anxiety as facilitative than the non-elites. In musical performance too, stage fright isn't just for novices—as we've seen, it touches most everyone. "Fears of collapsing onstage, forgetting the music, losing the place in the score, and similar concerns are all commonly reported by performers of many different levels of training and accomplishment," Salmon and Meyer write. "It's just that the professionals have learned that such fears are seldom truly predictive of what will happen during the performance. In other words, they treat such thoughts as a normal part of performing and are not unduly distracted or frightened by them."

With stage fright, acceptance is the better part of valor. The virtuoso classical pianist Arthur Rubinstein called performance anxiety "the price I pay for my wonderful life." Another pianist, André Watts, likewise described stage fright as "a fact of life," an entrenched yet benign character in his performances. What these top-level performers have figured out, after having engaged in enough fruitless hand-to-hand combat with their nerves, is that the thing that hurts performance isn't anxiety but *preoccupation* with anxiety—focusing on your fear rather than on the task at hand. The UCLA anxiety specialist Michelle Craske explains that the key to dealing with performance nerves is to learn to channel your energy away from self-monitoring, worry, and the whole anxious cavalcade, and send it back toward the job you're there to do. "In a performance situation, the methods for regulating anxiety are all about directing attention away from self-evaluation and onto the task," she said. "It helps you devalue the importance of the audience and place more emphasis on the story that's being told through the music." Keating would agree. "When you're focusing on perfection, you lose the whole essence of music," she said. "As soon as you stop being so focused on yourself, that's when you start being musical."

Because the symptoms of anxiety tug so insistently at our attention, psychologists advise performers to get used to handling distractions while playing. Salmon tells his charges to run up a flight of stairs or drink a mug of coffee (just like our friend Ogi Ogas) before sitting down to practice so they'll grow accustomed to performing with a thudding heart or quivering hands; if feeling nervous is routine, you won't feel the need to focus on it during a performance. Training with distractions under realistic conditions also works well. "The more you can approximate the circumstances of an actual performance, the greater the carryover is likely to be," Salmon said. Play a tape of an audience while you rehearse, he advises, or ask a friend to do something distracting as you run through a speech—loudly tap a pencil, say. This distraction-training method boasts one particularly successful adherent: Tiger Woods. When Woods was growing up, his father, Earl, liked to cough or rattle the coins in his pocket during Tiger's backswing in practice; Earl had a degree in psychology and believed, correctly, that this would toughen his son's mind.* Another tactic is to find something you can use on stage to divert your attention from your anxiety. For example, some public speaking experts advise nervous speechmakers to grip the podium so forcefully that it's painful. (Of course, if you inadvertently destroy the lectern in the process, that may raise uncomfortable questions.) The singer Carly Simon takes this pain-as-distraction tactic to its extreme; before a performance, she's been known to jam pins into her hand and even ask her backing band to spank her, all to draw her mind away from her nerves.

The lesson here is this: when we come to expect and accept anxiety as natural, we can begin to live in harmony with it instead of in opposition to it. As the longtime Canadian Olympic basketball coach Jack Donohue memorably put it, "It's not a case of getting rid of the butterflies. It's a question of getting them to fly in formation."

Grand Finale

One day in 1889, a young Indian lawyer rose from his seat in a Bombay courtroom and prepared to argue his very first case in front of a judge. The lawsuit at issue was trivial, yet the idea of

---

*Well, it worked when it came to *golf course* distractions, at least. We'll leave Tiger's more troublesome extracurricular distractions for another book.

speaking before the court filled him with dread. "I stood up, but my heart sank into my boots," he later wrote. "My head was reeling and I felt as though the whole court was doing likewise. I could think of no question to ask." Frozen and blank, he had to claim he wasn't fit to argue the case, and he hurried from the courtroom, humiliated. On another occasion, the lawyer wrote a small speech to read at a meeting of a local vegetarian society—hardly a vicious crowd. "I stood up to read it, but could not," he recounted. "My vision became blurred and I trembled, though the speech hardly covered a sheet of foolscap." He asked a friend to read the speech instead. Paralyzed by shyness and fear, the lawyer couldn't even manage to speak at a dinner party without being overcome by nerves. He nearly gave himself up as a lost cause.

This young man's name was Mohandas Karamchand Gandhi, though these days he's more commonly called *Mahatma* Gandhi, after the Sanskrit word for "great soul." His dreadful false starts at public speaking notwithstanding, we can safely say that Gandhi went on to gain a certain mastery of the form. But though Gandhi became a legendary speaker and spiritual leader, he never truly grew comfortable with the role; he often mentioned the "awful strain of public speaking" in his writing. What carried Gandhi through these ordeals was his tremendous moral courage: the causes he believed in meant far more to him than avoiding his fears. (This attitude, you'll recall, perfectly echoes Steven Hayes's acceptance and commitment therapy model.) Ultimately, the anxiety was unimportant. The meaning of what he was doing was what mattered. It was the same for Laurence Olivier, who forced himself on stage night after night despite severe nerves, out of dedication to his craft. And for Bill Russell, who never let a little locker room nausea deter him from his love of the game.

And, of course, it's true of Zoe Keating, who endured more than her share of strife before learning to interlace her anxiety with the music she loved. Now the same nerves that once tortured her just seem like old friends coming to visit. "It's normal for me to have a fight-or-flight response when I get up there on stage," Keating told me. "But I know it doesn't *mean* anything. I notice it now and it's like, *Oh, look at that, I'm nervous again*, and then I move on with what I have to do."

On a winter evening in San Diego not long ago, Keating found herself in what most musicians would call a nightmare scenario. Playing alone in front of a sold-out crowd of 1,500 as the opening act for the singer Imogen Heap, her electronic equipment suddenly failed. Half of the touring crew — including Keating — were feeling feverish and ill after a disastrous stop at a grocery store salad bar. The crowd seethed impatiently. At this point, even the most placid and cool-headed of entertainers might have hustled off stage with a quick "You've been great!" But Keating decided instead to play around a bit. "So I'm on stage in front of 1,500 people and my equipment isn't working, and I realize I'm not going to be able to figure it out," she recalled. "I said to them, 'Listen, I have no idea what's going on, so we're just going to improvise.' I'd been going for a minute or two when I reached this calm clarity I tend to get when I'm playing, and I realized what had gone wrong." Without skipping a beat, Keating leaned over and pressed a few buttons, bringing her electronics back online. She kept improvising, layering on new parts. "I created this whole piece on the fly that now I actually do for a lot of performances, pretty much exactly as it was on that day," she said. "I've already recorded it for my next album. It's called 'Don't Worry.'"

CHAPTER SIX

The Clutch Paradox: Why Athletes Excel — or Choke — Under Fire

As of the autumn of 1971, any conversation about clutch performance under the blazing floodlights of professional sports had to start with one man: Steve Blass.

Lantern-jawed and handsome, with an easygoing grin, Blass was the ace right-handed pitcher for the Pittsburgh Pirates, a rare talent whose career was cresting into the baseball stratosphere. Blass was what baseball insiders call a control pitcher. Instead of using flame-coated fastballs or physical intimidation to dominate, Blass succeeded because of his uncanny throwing precision; standing sixty feet away, he could fire the ball through a spot the size of a drink coaster, pitch after pitch. Under pressure, his accuracy only tightened. When the Pirates made the 1971 World Series—and immediately lost the first two games to the favored Baltimore Orioles—Blass turned the series around by throwing a masterful three-hitter in Game 3, spurring the Pirates to a 3–1 victory. Five days later, with the series knotted at three games apiece, the Pirates tapped Blass once again to pitch the deciding seventh game. Yet even faced with the gravity of baseball's highest stage, Blass maintained an astonishing air of good-natured nonchalance. In batting practice before that game, *The New Yorker's*

Roger Angell watched in amazement as Blass entertained his teammates with a parody of fellow Pirate Roberto Clemente's batting stance; "I had never seen such a spirited gesture in such a serious baseball setting," Angell wrote. A few hours later, Blass marched onto the mound and threw nine more flawless innings to give the Pirates the world championship. Just after he delivered the final out, an Associated Press photographer snapped a famous shot of Blass leaping ecstatically into the air, arms and legs splayed, a look of pure joy painted on his face.

At that moment, Blass seemed to embody a new form of clutchness in sports: not just rising to the occasion, but embracing it with childlike delight. "Nobody had more fun out there than I did," recalled Blass, who now works as a commentator on Pirates broadcasts. "It was a hoot for me. It wasn't a career, it wasn't a job—it was fun." The following year, his enthusiasm for baseball translated into even greater success. Blass won his first selection to the National League All-Star team. He appeared on the cover of *Sports Illustrated*. With a stellar earned run average of 2.49 to go with his nineteen wins, Blass was putting up the kinds of statistics that would win a pitcher a nice $100 million contract in today's game. (But since this was *yesterday's* game, he pulled in around $90,000 per year instead.) Blass was a flesh-and-blood baseball hero.

Now fast-forward to the spring of 1975.

Less than three years after hitting the apex of his sport, still in the middle of his physical prime, Steve Blass had fallen so far out of the major leagues that he was peddling class rings as a traveling salesman for Jostens. His pitching career had cratered, but not because of injury, personal tragedy, or scandal. Instead, in one of the greatest enigmas in baseball history, Blass simply lost the ability to keep his cool during games. He hadn't gone crazy; off the field, Blass still seemed as unruffled as ever. And his throwing arm had never felt better—in fact, he could still hurl perfect strikes to the catcher at will during practice. But whenever Blass stepped onto the mound and saw a batter hunched over home plate, performance anxiety triggered some mysterious physiological glitch that transformed him into a different pitcher entirely: an errant-throwing amateur. Suddenly he had to struggle just to get the ball into the catcher's mitt. (He even started throwing the ball *behind* batters.) His joy for baseball dissolved into dread; Blass now prayed not to be sent into games. When the Pirates sent him down to the minor leagues to recover, the recent All-Star became the worst pitcher, statistically, in his entire league. Soon enough, the same guy who was cracking wise with his teammates right before Game 7 of the World Series couldn't sleep the night before he had to throw in batting practice—to other pitchers...*minor-league* pitchers. On the baseball diamond, Blass no longer could handle any pressure *at all*.

Of course, every athlete endures occasional slumps, but Blass's throwing breakdown was unprecedented: it struck without warning; it caused a dizzyingly steep drop in performance; and it appeared to be completely untreatable. Mystified baseball insiders began whispering about "Steve Blass disease," and no one was more baffled than Blass himself. "I kept asking myself, 'Why the hell is this happening?'" Blass told me from his Pittsburgh home. "I was in the prime of my career and I had the kind of arm where I could have pitched until I was forty. But there was this apprehension about throwing, when I got ready to cut the ball loose. It wasn't free and easy anymore." In 1974, sitting in a Florida restaurant before a spring training practice, Blass explained to a *Sports Illustrated* writer how his throwing anxiety felt. "It never leaves me," Blass said. "Before, I'd just go out and do it. I never thought about it....Now, it scares me. Scares the hell out of me....You don't know where you're going to throw the ball. You're afraid you might hurt someone. You know you're embarrassing yourself but you can't do anything about it. You're helpless." After two years of struggle, Blass quit his sport for good. For decades, his fear even kept him from being able to toss around a baseball in his backyard.

When we try to label the psychic malady that befell Steve Blass, this is the first thought that springs to mind: on a truly epic scale, he *choked*. At least technically, it's the truth. As the researchers Sian Beilock and Rob Gray define it, choking under pressure is "suboptimal performance—worse performance than expected given what a performer is capable of doing and what this performer has achieved in the past." Blass clearly had the ability to accomplish incredible, *clutch* feats on the pitching mound. Yet now, for reasons no one could fathom, Blass was underperforming (that is, choking) on every single pitch. He felt anxiety, sure—but so did plenty of other athletes and performers, like Bill Russell or Laurence Olivier, who managed to succeed without a visible hitch. Blass's choking problem was a nervous one, yet it went deeper than the performance nerves we explored in the last chapter. Unlike stage fright, which springs from an *external* force (the audience), Blass's issue seemed to stem from some nebulous *internal* glitch. Plus, Blass still choked with virtually no one watching him—so performance anxiety alone couldn't explain it. It was as if Blass had split into two people, the poised sportsman who had thrown pinpoint sliders before an audience of millions, and the anxiety-plagued amateur whose arm refused to work properly. Each version of Blass had the same body, the same talent, and the same store of baseball expertise. Only one thing was different: his mind. Some hidden psychological shift turned a World Series hero into a class ring salesman.

In the universe of sports, there are few dirtier labels you can apply to someone than "choker." (Two of them would be "referee" and "baseball agent Scott Boras.") It's a reputation that an athlete can earn in an instant and never live down. In 1991, with eight seconds left in Super Bowl XXV, the Buffalo Bills placekicker Scott Norwood—a solid, Pro Bowl–caliber player—tried a forty-seven-yard field goal to defeat the New York Giants and missed wide right, losing the Bills the game. Dubbed a choker, Norwood was soon out of football; he now sells real estate. In Game 6 of the 1986 World Series, the Boston Red Sox first baseman Bill Buckner—likewise a solid player and a former All-Star—attempted to field a ground ball with two outs in the tenth inning to lock up the championship against the New York Mets, but the ball rolled under his glove and through his legs. The Mets won the series, and Buckner became the most loathed goat in Boston sports.* When Buckner dies, his split-second World Series choke will be in the first line of his obituary, eclipsing everything else he ever accomplished. But while Norwood and Buckner will wear the "choker" label like a scarlet letter for eternity, so-called clutch athletes win a special kind of popular esteem. Indeed, if *choke* is a curse word, then *clutch* is its lofty polar opposite; coming through under pressure is what turns a mortal athlete into a legend. We admire the athletic talent of sports figures like Michael Jordan, Tom Brady, and Derek Jeter, but we *revere* them because they excel when the stakes are highest.

Despite the colossal emphasis that athletes and fans place on pressure performance, however, choking and clutchness remain poorly understood psychological phenomena in the sports world. For all of the scorn we heap on chokers and the acclaim we bestow on clutch performers, few of us—both inside and outside of

---

*Boston residents even nicknamed the Leonard P. Zakim Bunker Hill Bridge—which is shaped like an upside-down *Y*—the "Bill Buckner bridge," because so much traffic rolls between the structure's "legs."

organized athletics—have any idea what makes the two groups so different. Commentators speak vaguely about the importance of "mental toughness," but this stale term tells us nothing concrete about how heroes like Jordan and Jeter pull through in anxiety-packed moments. We're mostly in the dark, and no one on earth represents the enigma of choking and clutch quite as well as Blass, the man who has both soared to the pinnacle of performance under fire and crashed into its lowliest gutter. To understand how choking and clutch really work, we must first solve one mystery: What the hell happened to Steve Blass?

"The Creature"

Blass might have his name attached to the dreaded "disease," but plenty of other ballplayers have caught it as well—and when the psychic virus hits, it's tough to hide it. Take the case of Mackey Sasser, who was a promising young catcher for the Mets until he suffered a jarring collision at home plate in one July 1991 game. Immediately thereafter, for reasons he couldn't explain, the simple act of tossing the baseball back to the pitcher's mound suddenly became a comic dance of hesitations and double-pumps, culminating in a weak throw. Sasser's arm was fine—he could still fire a bullet to second base to gun down a base-stealer—but his mental block was severe enough that players actually stole bases while he fought to chuck the ball to the pitcher. His career tanked because of it. The New York Yankees second baseman Chuck Knoblauch developed the "disease" in 1999, when he irrevocably lost the ability to throw the ball accurately to first base. In one game, Knoblauch flung so far off target that the ball careened into the stands and nailed the broadcaster Keith Olbermann's mother in the head. Marie Olbermann fully recovered, but Knoblauch never did.

Steve Sax, another famous sufferer who eventually overcame the feared ailment, told me that these nervous performance glitches are more agonizing for athletes than fans would ever guess. "It was horrible," he said of his Knoblauch-like throwing troubles in the 1983 season, when he was a second baseman for the Los Angeles Dodgers. "It was the worst thing I ever went through besides losing my parents. I thought I was going to lose my career." Sax felt like the universe was actively conspiring to make his troubles worse. At games, fans sitting behind first base would sarcastically don batting helmets for protection, or hold up signs saying SAX — THROW ME A SOUVENIR! "I remember one guy in Atlanta right above the dugout had worn this prosthetic arm, and when I'd come off the field he'd flap it all over the place," Sax said. His manager, Tommy Lasorda, offered little in the way of compassion either. Once, Lasorda—who wasn't exactly the touchy-feely type—chided him, "Do you know how many people can throw a baseball from second base? A billion." Not so helpful. Sax even received threats from shadowy sports betters who believed his errant throws were messing up the spread. It's tough to recover your stride, Sax said, when your self-confidence is being assaulted on all fronts.

The anxious blips of flatlined performance we see in those like Sasser and Sax may seem peculiar, but the psychologist and former ballplayer Tom House says they're more common than we think. House, who had a low-wattage career as a major-league pitcher in the 1970s (his best-known feat was catching Hank Aaron's record-breaking 715th home run out in the bullpen), has counseled dozens of players on the issue, in cases both massive and slight. House calls the mental block "the Creature." Whenever fans make dismissive comments about a professional player choking, House has to chuckle at their naïveté. "If I asked you right now to stand up, walk over, and touch the closet door, then go sit down again, you wouldn't have to think about it," he said.

"But if I told you that if you didn't do it right, there would be 50,000 people booing and you'd have to see it replayed on Sports-Center and you might get released from the team, then it's not so simple." Often, when afflicted players must react quickly (like Sasser throwing out a base-stealer), they surprise themselves with their accuracy. It's only when they have time to *think* that the Creature emerges. And, of course, a pitcher like Steve Blass has time to think before every single throw.

Even now, nearly four decades after his precipitous fall, Blass has little insight into what caused his anxious saga. "People had all sorts of theories—that I got to a stressful point where I was *expected* to be good, or that I was affected by Roberto Clemente's death [in a 1972 plane crash]—but I never bought into them," said Blass, who is now in his late sixties. "To this day, I really don't know what it was." His decline was doubly confusing because from his earliest playing days growing up in tiny Falls Village, Connecticut, Blass had felt nothing but pure joy for baseball. Even after he arrived in the majors in 1960, he refused to make the game into a serious, humorless job. "We used to have a prize day at the end of the season," Blass recalled, "and between the innings they'd drive in a car that they were going to give away, right behind home plate. They had to stop doing it, because I would try to knock the windows out of it when I was warming up. I mean, I always had a sense of fun about it."

It's not as if Blass never got anxious on the mound. Indeed, he even felt nervous in his clutch World Series performance—at the beginning. "I walked out there in the first inning with sixty million people watching, and I said to myself, 'I'm from a town of eight hundred people in Connecticut—what am I doing in the seventh game of the World Series?'" he told me. Blass immediately felt ill at ease, but soon the Orioles manager Earl Weaver trotted from the dugout to complain to the umpire that Blass had violated an arcane rule stating that pitchers must throw from the front of the rubber. "It was just one of those nickel psychological ploys to disrupt me, but it got me so pissed off that I locked back in and found my rhythm again," Blass said. Such tense moments were rare for Blass, however. In general, he maintained a happy and calm demeanor, and his teammates prized him for it.

If Blass was such a laid-back guy, though, then what could have possibly plunged him so abruptly into constant choking? One very, very bad game. At the beginning of the 1973 season, fresh off his All-Star year, Blass hit a normal-seeming slump. No one worried too much about it; the Pirates coaching staff worked on tweaking his throwing mechanics and assumed he would soon emerge from his rut. Then one mid-June day, in a run-of-the-mill game against the Atlanta Braves, Blass just *cracked*. With the Pirates leading 8–3 in the fifth inning, the manager Bill Virdon called him in for a few confidence-building innings, but the Steve Blass who jogged onto the mound that day was no longer the same precise pitcher as before. Looking uncharacteristically tense, Blass found himself unable to deliver the ball accurately. He flung it into the dirt, threw behind batters, and lobbed softballs that the Braves smacked around with ease. In barely more than an inning, Blass allowed a whopping seven runs. "It was the worst experience of my baseball life," he later told *The New Yorker*'s Angell. "I was totally unnerved. You can't imagine the feeling that you suddenly have no *idea* what you're doing out there."

Over the next few months the Pirates continued to play him, hoping he'd buck up, yet Blass had transformed into a different athlete—an anxious, avoidant *amateur*. As one of Blass's friends put it to Angell, "In the old days, he used to get mad at himself after a bad showing, and sometimes he threw things around in the clubhouse. But after this began, when he was taken out of a game he only gave the impression that he was happy to be out of there." Even Blass's witty demeanor evaporated. Blass soldiered on in hopes of recovering, but the experience felt like torture; it took all of the courage he had just to stand on the mound. In a September game against the Chicago Cubs, Blass pitched five decent innings and for one glimmering moment believed he'd snapped out of his slump. (He even called his wife and yelled, "I'm back! I'm back!") But that game would be his last passable pitching performance. He ended the 1973 season with a 9.81 ERA, worst in the National League. And in the minors the following season, things fared even worse.

So what, exactly, had changed within Blass? What was different when he tried to throw the ball now? Blass knew what it felt like to play when nervous, but his problem went beyond anxiety; it was as if an evil imp were hiding inside his body and sabotaging every throw. "I could start the windup the way I wanted to," Blass explained, "but then at the point of release it just *froze*. Way down deep, I'd say to myself, *This is not going to work*." Pitching, it should be pointed out, is an extraordinarily delicate and precise task; if a pitcher's release point is off by a few thousandths of a second, the ball might sail over the batter's head. Now Blass felt out of sync like this on every throw. On the mound, he fidgeted and grimaced. In 1974, the *Sports Illustrated* writer Pat Jordan sketched a vivid portrait of his throwing discombobulation: "He raises both hands overhead, and suddenly his right leg, the one in contact with the rubber, begins to wobble uncontrollably. From a distance that leg looks as if it has the consistency of an overcooked strand of spaghetti. Up close, it looks as if that leg is expressing an urge to flee."

Blass tried everything he could think of to get back on track. He tinkered with his throwing motion. He practiced pitching from the outfield, or while kneeling. He pitched perfect mock games in which his fastball shot straight to the catcher's mitt and his slider nicked the edge of the plate—and it all fell apart whenever a batter stood in. Blass went even further, trying transcendental meditation. "I went to a hypnotist, too, and they said I wasn't a very good subject—I was too strong-headed." Blass laughed. At the apex of his troubles, Blass received fifty letters per week from would-be psychoanalysts; his favorite was from a Virginia hunter who claimed to have discovered that whenever his aim was off, his underwear was too tight. (Blass roared with laughter, paraded the letter around for his teammates, then rushed out to buy looser underwear. They didn't help.)

The harder Blass struggled against his throwing anxiety, the worse things got. And the press's interest in the curious case of the plummeting pitcher just intensified his misery; everywhere his minor-league team traveled, someone would ask Blass if he thought he still merited his salary. It was a grueling experience that tested his character. "The thing I was proud of was that I had time for everybody and I didn't become an ass about it," Blass told me. "My philosophy was that if I had time for 'em when it was going good, I should have time for 'em now that it was going bad." He tried to remember that he still had a happy life with his wife, Karen, and their two sons. Blass reported to spring training at Bradenton, Florida, in 1975, but in exhibition games his pitches were still woefully off target. Fans began booing him. So one day, after an encouraging performance in practice during which he struck out a few batters, Blass decided to go out on a high note: he walked out of the clubhouse, hopped into his Volkswagen, and drove back to Pittsburgh. His playing career, he knew, was over.

But if Blass had encountered the Creature today instead of thirty-five years ago, things might have turned out differently. Recently, psychologists have compiled a mountain of new research on choking in athletics and skill-based tasks—and as any of them could tell you, Blass put himself on a collision course with disaster from the very moment his problem first struck.

Don't Look In!

On the morning of February 8, 1986, as a group of basketball sharpshooters practiced for that evening's NBA three-point contest in Dallas's Reunion Arena, the great Boston Celtic Larry Bird started a seemingly offhand conversation with a young player named Leon Wood. A second-year shooting guard with the Washington Bullets, Wood was the favorite at the competition; while practicing the week before, he had hit twenty-eight of thirty-one three-point shots in one uncanny stretch. Larry Bird was...well, Larry Bird—a legendary clutch player at the apex of his powers, with a legendary ego to match.

"Leon," said Bird, as overheard by the *Sports Illustrated* writer Jack McCallum, "I've been watching you. Are you shooting different than you used to?"

"I don't think so," replied a suddenly perplexed Wood. "Why?"

"I don't know," Bird said. "Something looks different about your release. Don't worry about it, though."

And then Bird trotted off, leaving the gears of Wood's mind to start spinning away. To make matters worse, shortly before the competition began, Bird strolled into the tiny room where the other seven competitors were waiting and informed them that the best they could hope for was second place. Wood grew so upset at this that he left the room. By the time the contest got under way, Wood was clearly off his game; he finished in seventh place. Bird, on the other hand, looked tranquil and unhurried as he nailed shot after shot through three rounds, winning first place, a $10,000 prize, and a free pass to run his mouth. ("I'm the king of the three-point shooters," he said afterward. "I always thought I was, and I proved it.") With just a few casual words to

Wood—plus a healthy dose of bravado—Bird had made his main rival choke. But how?

Over the last decade, Rob Gray, a professor of applied psychology at Arizona State University, has become one of the nation's leading experts on what makes athletes choke under pressure. Gray's primary research tool is a virtual batting cage, where he sends experienced collegiate baseball players through a variety of scenarios to see where and why their performance falls apart. "It's basically set up like a large Wii video game, where people interact with a simulation of a pitcher in the field," he explained. "There's a real bat, and the subjects get auditory feedback and see the balls fly out on the video screen. We use motion tracking equipment as well, so we can measure not just the movement of the bat but the movement of their limbs." Even with elite Division I baseball players who are used to tense game situations, Gray can introduce stressors that make their batting cage performance drop by 30 percent or more—say, sparking social pressure by telling a player that his team will get a cash reward only if *he* gets a certain number of hits. This much, at least, isn't too surprising.

What *is* surprising, however, is the mechanism that makes them choke: not anxiety, but simply paying greater attention to their swing. In a 2004 study, for example, Gray gave batters two different tasks. First, he played a tone during their swing and asked them to note whether it was high or low in pitch. This task had no effect on their performance. But then Gray asked the subjects to focus on whether the bat was moving upward or downward when the tone played. "Basically, we tried to get them to focus their attention inward, which isn't what they usually do," Gray said. "And when they did that, it hurt them really badly." Because they've rehearsed it so many times over the years, high-level players can typically execute a perfect swing without giving it a thought. Gray's motion trackers confirmed this: under zero pressure, they each stroked the bat smoothly and consistently. Yet when he had them pay extra attention to the process of swinging, the players suddenly showed an unpredictable jerkiness that fit the same pattern as choking. Somehow, the internal focus threw them off—and this is the weapon that Bird used on Wood. By needling the inexperienced Wood into analyzing his shooting mechanics (and, of course, Bird hadn't *actually* seen anything different in his release), Bird all but ensured that the young guard would blow it under pressure.

Appropriately, this hypothesis about the roots of choking—called "explicit monitoring theory"—originated with a blunder. During Super Bowl XIII in January 1979, the psychologist Roy Baumeister watched the Dallas Cowboys tight end Jackie Smith inexplicably drop a touchdown pass that hit him right in the hands. The mistake cost the Cowboys the game. In Smith's choke, Baumeister saw a psychological paradox. "I was thinking, *He'd catch that ball ninety-nine out of a hundred times in practice, so what's the deal?*" he told me. "Why, in the biggest moment in his life, does he drop it? In psychology at the time, the assumption was that the harder you tried, the more important something was, the better you did. But at times like this, things went in the opposite direction: the importance of doing well made people do worse." Baumeister theorized that in high-tension moments our natural response is to pay hyperclose attention to what we're doing in order to increase our odds of success, yet this attention somehow disrupts us. Soon, studies began bearing his idea out. In one seminal 1984 experiment, Baumeister found that telling subjects to concentrate on their hands as they did a manual dexterity test made them perform far worse than if they were given no instructions. One study even found that simply mentioning that "some people have the tendency to choke at the free-throw line" was enough to make inexperienced free-throw shooters significantly more self-conscious and error prone.

And why is inward attention so toxic to performance under pressure in skill-based tasks like sports? Here's the simple explanation: it instantly turns experts into novices. Think of learning to ride a bike. At first, it's an arduous and deliberate process. You have to *concentrate* on doing it correctly. As you gain experience, you begin thinking about the process less and less, until bike riding one day becomes *automatic*. Your expert bike-piloting knowledge, then, doesn't live in the conscious mind. It lives instead in the unconscious departments of the brain that manage motor control (like the cerebellum), enabling you to pedal smoothly without thought. The exact same process is at play in skill-based sports. Athletes take a split-second physical act (swinging a golf club, pitching a baseball) and refine it through thousands of careful repetitions, until the motion becomes automatic and unconscious. Once an athlete has perfected his mechanics, things flow best when he doesn't think about them, because the procedures for ideal performance are now completely internalized.

When anxiety turns our attention inward and we try to consciously control a highly rehearsed skill, we essentially ditch our hard-won subconscious expertise. "If self-awareness makes you take something automatic and suddenly start paying attention to it again, that disrupts the whole process," Baumeister explained. "It's shifted you from a skill execution mode back to a learning mode where you're figuring out what to do—only the conscious mind doesn't know how to do it." Psychologists call this process "dechunking," and Gray likens it to taking a superefficient Porsche engine and randomly tweaking some bolts; you're just going to introduce problems. As he and Beilock have written, inward attention under pressure breaks a seamless motion like a baseball swing "back down into a sequence of smaller, independent units, similar to how the performance was organized early in learning," and each of these units becomes newly susceptible to failure. Internal monitoring simply turns expert performers into jerky, hesitant, mistake-prone novices—like a certain Pittsburgh Pirates pitcher with whom we're acquainted.

Applying more vigor to *effort*-based tasks like sprinting or weightlifting can actually boost performance, but with tightly honed *skills* like pitching, it only causes trouble.* This monitoring-choking connection has been established in study after study. Rock climbers grow more deliberate and rigid in their movements when high up on a climbing wall than when low to the ground; they become more conscious of the greater stakes of falling, which makes them focus inward instead of on the task at hand. Experienced soccer players dribble the ball worse when asked to concentrate on the side of their foot, but not when simply distracted by noise. The sport psychologist Bradley Hatfield says that when you examine brain scans of athletes about to choke, their minds look like a "traffic jam" of worry and self-monitoring, whereas the brains of those who do well under stress appear efficient and streamlined, engaging fewer neural regions. "Simple mind, consistent performance," Hatfield told me. "Complex mind, greater variability in the performance."

Adventurous readers who want to test out explicit monitoring theory can easily do so from the comfort of their own homes: just try *not* to do something. Earlier, we encountered the Harvard psychologist Daniel Wegner, whose research showed that asking subjects not to think of a white bear actually made them *more* likely to have white-bear-related thoughts. Well, the same trick applies to

---

*This fact has led Baumeister to an intriguing pet theory. "You know how they say offense sells tickets and defense wins championships?" he asked me. "Well, I think a good reason for that is that in basketball or football, defense works mainly through effort while offense is more about skill." In other words, in a pressure-packed championship game, the offense should be more prone to choking than the defense.

physical tasks; our attempts to prevent an undesired outcome often cause that very thing to happen. If you've ever traversed a white carpet with an overly full mug of coffee, murmuring, *"Don't spill, don't spill,"* only to watch several large brown drops plummet below, you know exactly how this works. Ask someone to hold a pendulum steady, as Wegner has done in his lab, and they do a fairly stable job. Ask them *not to move it side to side,* however, and the pendulum magically starts meandering left and right. When we issue internal orders like "don't spill the coffee," Wegner explains, the brain likely just ends up activating its "spill the coffee" circuitry, producing what he calls "ironic effects." The upshot of this for an athlete is obvious: attempting *not* to throw a wild pitch or brick a free throw can make these things more likely to occur.

Add all of this choking research up—the perils of monitoring well-honed mechanics, trying harder under pressure, and attempting *not* to screw up—and what do you get? You get Steve Blass. Blass's issues, remember, began with an ordinary slump. Yet as Gray explains, the real trouble starts when an athlete responds to natural performance lulls with unnecessary tinkering. "Once a player hits a statistical trough of bad performance, he ends up making it worse when he starts doing things to change it instead of letting it pass," he told me. And this is precisely what Blass and the Pirates did: they went under the hood with Blass's throwing motion and "fixed" what wasn't broken. In reality, there was nothing wrong with Blass mechanically, just as there was nothing wrong with Sasser or Knoblauch. Their issue was overthinking, and the more they tried to correct the illusory mechanical "problem," the more inward focus they developed. Blass readily admits that his efforts to modify his pitching motion merely burdened his mind with worries. "When you're pitching your best, there's no clutter," Blass said. "You don't have to think about mechanics.

Everything is slow and nice and fluid. But when you get absorbed in all of that *stuff*, there's a lot of clutter. You're tight, not relaxed." Hence the old baseball adage, "Thinking is stinking."

The true source of Steve Blass's "disease" was this: performance anxiety made him overthink his pitching motion, and this internal attention morphed him into a wobbly amateur. It's a problem that couldn't be fixed with tweaks or greater effort or even looser underwear. Blass-style overthinking is the thread we see running through all blockbuster choking victims. The golfer Greg Norman, for example, held his sport's top ranking through much of the 1980s and '90s despite an almost mythological tendency to collapse in big moments. At the 1996 Masters, Norman was up by six strokes on Saturday, yet he sprayed the ball all over the course in Sunday's final round, eventually losing by five strokes in one of history's epic chokes. "The more I tried, the more it went away," he later wrote of his performance. On day four of those tournaments, Norman essentially became Steve Blass—an expert sunk to novice level by internal attention. The sport psychologist Shane Murphy even has proof of this: after Norman combusted ten years earlier in the 1986 Masters, Murphy surveyed the video of Norman's shots from all four days, picking up on one key difference. "His pre-shot routine on the last day was eight or nine seconds longer than his pre-shot routine on the first day," Murphy told me. "Something was clearly unsettling him, and it looked like overthinking."

Psychologists have formulated a number of methods for overcoming the tendency to choke, most of which revolve around diverting an athlete's attention from anxious internal monitoring. "What works is if you get something else to focus on, like your opponent or the conditions," Baumeister told me. "When they say 'Keep your eye on the ball,' yeah, it helps to focus on the ball, but it also keeps your attention off yourself." Ideally, Gray says, athletes will concentrate on cues that are relevant to what they're doing—a batter might zero in on the movements of the pitcher, for example—but *any* distraction will do. "Even if they're paying attention to the music that's playing, it prevents them from interfering with the action itself," Gray said.* For some, concentrating on a cue word like *smooth* provides a useful distraction. And virtually every sport psychologist who has ever donned a polo shirt advises athletes to focus only on the present moment's task— delivering *this* specific pitch, shooting *this* next free throw— instead of slipping into future-focused worries about what's at stake, or what the results of the game will be.

But often, struggling players resist anything resembling therapy. "Athletes have a strong sense of who they are, and the thought of *I'm not messed up* is very important," explained Ken Ravizza, who has counseled players on several major-league baseball clubs. "We run into that barrier of the shrink image." So players frequently turn to another time-honored solution: becoming thoroughly obsessive-compulsive.

When You Believe in Things That You Don't Understand...

Take a look at a professional baseball game—with the players all swaggering and spitting and adjusting their nether regions—and you might believe you're watching some of the most macho, fear-

---

*The Hall of Fame San Francisco 49ers quarterback Joe Montana used this distraction method to great effect in Super Bowl XXIII. With his team down by three, time running low, and his teammates nerve-racked and edgy, Montana scanned the crowd and exclaimed to the players in the huddle, "Look, there's John Candy!" The humorous external focus loosened them up, and they marched down the field for the game-winning touchdown.

resistant men on earth. But examine the on-field scene more closely, and you'll notice something peculiar: these athletes seem to repeat the exact same motions and tics before *every play*. Here's Player A muttering an incantation to himself before each pitch. There's Player B readjusting his batting gloves and then clapping three times whenever he steps away from the plate. Suddenly this manly game starts looking more like a cavalcade of rituals and superstitions designed to placate players' nerves. Stout men with arms like tree trunks live in terror of stepping on the chalk foul line as they trot to the dugout. Others swear that each bat can only produce a certain number of hits before becoming useless; for Honus Wagner, that number was one hundred hits per bat, after which he'd discard the bat forever. Many athletes contend that rituals like these are innocuous—an argument that falls apart the moment you request that they *not* perform them. When asked what would happen if he didn't adhere to his in-game routines, the pitcher Yorkis Perez replied matter-of-factly, "I'll get sick or die." This, one psychologist explained to the *Seattle Times*, is "the exact same mechanism, clinically, we find in obsessive-compulsive people."

No ballplayer in the modern era has exemplified the neurotic addiction to superstition more than the great hitter Wade Boggs, whose inflexible daily habits numbered in the hundreds. As outlined in one 1986 *Sports Illustrated* story by Peter Gammons, Boggs's game day routine began at two p.m. with a chicken dinner. (As a rookie, Boggs noticed a correlation between eating chicken and multihit games, and from then on he had his wife fix him chicken every day, on a rigid thirteen-recipe rotation.)* Boggs always departed from his apartment at three p.m. sharp, arrived at the clubhouse at three fifteen, and sat in front of his locker at

---

*It was actually a fourteen-day schedule, but Boggs had such faith in the efficacy of lemon chicken that he requested it once a week.

three thirty. He took the exact same number of practice ground balls at third base each day. Maybe none of this seems strange so far, but wait. After warming up in the field, Gammons writes, Boggs "steps in order on the third, second and first base bags, steps on the baseline...takes two steps in the coach's box and lopes to the dugout in exactly four steps. Because he always goes to the first base dugout via those four steps, they are clearly visible on the Fenway [Park] sod by August." Various other rituals filled the next few hours, until at precisely 7:17 p.m. Boggs jogged out to do pregame wind sprints. (An opposing manager once arranged to have the game clock flip directly from 7:16 to 7:18 in an effort to throw Boggs off; it worked.) In the batter's box, he always sketched the Hebrew word *Chai* in the dirt with his bat—somewhat mysteriously, since Boggs wasn't Jewish. All the great players have hitting routines, Boggs told Gammons: "Mine's the same as theirs—only it takes a little more than five and a half hours."

Why, we might ask, do ballplayers embrace these rituals despite all reason and good sense? They do it because their superstitions give them a sense of *control* and help reduce the *uncertainty* of how they'll perform, both of which diminish their performance anxiety. At its heart, superstition is simply a tool to help us manage fear: if we believe that eating chicken will tip the future in our favor, we'll be less anxious about it, even if there's no actual cause-and-effect relationship. In his entertaining paper on "Baseball Magic," the anthropologist and former minor-league ballplayer George Gmelch draws an intriguing comparison between how pro athletes approach baseball and how the natives of the Trobriand Islands, off New Guinea, approach fishing. The Trobrianders fish in two spots: the safe inner lagoon, and the perilous open sea. Within the secure confines of the lagoon, Gmelch explains, "where men could rely solely on their knowledge and skill," the fishermen use virtually no rituals. Yet amid the danger and uncertainty of the open sea, the Trobrianders employ superstitions constantly to give themselves a feeling of control over their fate.

As Gmelch points out, we see the same pattern in baseball players. Almost no one has rituals for *fielding*, an unthinking reflexive reaction that major leaguers perform correctly nearly every time. Yet virtually everyone has rituals for *hitting*, because even the best players will fail to reach base in most of their at bats. Hitting is unpredictable, and thus more anxiety-provoking; superstitions give players a feeling of control, curbing that fear. (As a general rule, people with the least amount of predictability in their lives tend to be the most superstitious—think of gamblers and performing artists.) We might scoff at Boggs's rituals, but hey, the man maintained an astronomical .328 career batting average. If he believed these things worked, then maybe they did, just not for the reasons he thought. Problems only arise when we, like those with obsessive-compulsive disorder, become *prisoners* of these routines, which is why you won't hear many psychologists advising people to ease their anxieties by adopting new superstitions.

As an example of a more levelheaded deployment of athletic rituals, consider the NFL kicker Adam Vinatieri, the most celebrated clutch player in the history of his position—a man whose last-second field goals won the New England Patriots two Super Bowls. In a 2007 *New York Times Magazine* profile, Vinatieri revealed the secret of his success to the writer Michael Lewis: he performs every single kick with the same machine-like precision, removing his brain from the equation entirely. Opposing coaches often try to "ice" kickers before a field goal by calling a time-out, giving them time to think (and then, hopefully, choke), but by factoring out as much mental activity as possible, Vinatieri hopes to become ice-proof. On each try, Vinatieri zeroes in on the spot where the holder will place the ball, then takes two painstaking steps back and two to the left, never moving his eyes from his mark. He waits, still and thoughtless, for the holder's hand to move. Then, with the same cadence each time, it's *step, step, kick*, the process taking between 1.3 and 1.5 seconds. "He arranges his kicking routine to prevent his mind from playing any role at all," Lewis writes. On the foundation of this robotically consistent, choke-resistant procedure, Vinatieri has built a legendary career. It's important to note, though, that Vinatieri's routine is quite different from a superstition, because everything he does directly affects the task at hand. While Boggs's rituals served no concrete purpose other than to lend him an artificial feeling of control, Vinatieri's kicking process is simply a tool to keep his mind clutter-free and his performance steady.

Solid routines like Vinatieri's are vital to success under fire, but his isn't the only way. Another useful routine many athletes employ to help them perform under pressure is mental visualization, a practice that long carried a fuzzy New Age vibe, until a stack of hard evidence for its efficacy thudded onto psychologists' desks. The man arguably most responsible for popularizing visualization was the legendary golfer Jack Nicklaus, who claimed that a good golf shot is 10 percent swing, 40 percent stance, and 50 percent pre-shot imagery—picturing his swing, the arc of the ball, and even the thump of the ball landing on the green grass. Lindsey Vonn, the gold-medal-winning American Olympic downhill skier, even takes this process beyond the mental realm: in preparing for a race, she'll balance her feet on two slacklines, close her eyes, and mime skiing the entire course, turn by turn. Visualization routines like these facilitate performance under pressure for an intriguing reason, says Shane Murphy. Research suggests that visualizing, say, a good golf shot constitutes a sort of neural practice for actually *hitting* that shot. "If you imagine a successful golf swing, you'll use nearly the same pathways as when you take the swing," Murphy told me. Thus, in a sense, visualization primes the mind not to choke.

All of these tactics—from external distractions to the establishment of success-enabling routines, from visualizations to an unshakable belief in the talismanic power of lemon chicken—can help an athlete avoid choking when the pressure is on. But another equally significant question remains wide open: How can someone *raise* his ability to meet the needs of a big moment? How, in other words, does someone become *clutch?*

A Tale of Two Pitchers

In a centuries-old sport like baseball, it's rare to see one athlete single-handedly change the way the game is played—but then again, the pitcher Dennis Eckersley was a rare character. Snarling and intimidating, with a horseshoe mustache above his lip and a long black mane flowing from beneath his cap, Eckersley looked like a pirate who had stormed into a baseball clubhouse and donned an Oakland A's uniform. Throughout his career, which reached its peak in the late 1980s and early '90s, Eckersley's cockiness both amazed and cowed batters. When Eck struck out a hitter, as he did at an astounding clip, he'd point at them menacingly, or shout "You're next!" at the player on deck. He entered games with "Bad to the Bone" blaring through the stadium. Eckersley backed these dramatic flourishes up with consistently dominant clutch performances. His job was to close out baseball games—to take the mound in the ninth inning and shut the other team down—and thus, every time Eckersley entered a game, it was a pressure situation. To say he excelled in this role would be understating it; in the 1990 season, for example, Eckersley had an infinitesimal 0.61 ERA, meaning he allowed less than one run on average per nine innings pitched. His control was so tight that some catchers said they'd be willing to hold up a mitt and catch his pitches with their eyes closed. Soon enough, every team wanted an intimidating, clutch late-game pitcher like Eck, and the role of "closer" was born.

So whenever this brash buccaneer of a pitcher stepped onto the mound for one of his high-stakes performances, one would assume he felt nothing but pride and naked aggression—certainly not *fear*. This is exactly what Eckersley wanted people to think. Here, though, is what he actually felt. "Basically, I'm scared to death out there," Eckersley said in 1992, at the zenith of his fame. "Every time I go out there, the fear is running through my head. What if I fail?" Behind the scenes, when he spoke candidly with reporters in the clubhouse, Dennis Eckersley became quite possibly the most openly anxious athlete of the modern era—as anxious as Blass, even. His cocky persona was "an act" to mask how terrified he felt, Eckersley admitted; if hitters thought he was half crazy, they'd never guess he was so afraid. Every game, waiting in the bullpen for his cue to warm up, he seethed with dread. On the uncommon occasion that he blew a game, Eckersley would stay awake all night, churning over what he did wrong.

Yet despite his pervasive anxiety, Eckersley remained a determined and impressively resilient ballplayer. He had plenty of opportunities to cave in to fear, after all. In the 1988 World Series, Eckersley gave up one of the most famous game-winning home runs in baseball history to the Los Angeles Dodgers' Kirk Gibson, which ultimately led to a Dodgers championship. This alone could have buried him in a Blass-style funk. Instead, he replayed his moment of failure for an hour after the game with reporters and came back a better player the next season, helping propel the A's to a World Series win. Naturally, all athletes get nervous—the elite ones, as we've seen, just tend to view their butterflies as facilitative to performance—but Eckersley provides us with an extreme and almost paradoxical case. He was, at once, one of base-

ball's most terrified players and one of its greatest clutch performers. So if Eckersley wasn't calm on the field, how on earth did he
manage to be so clutch?

Please steady yourself, because I'm about to impart a momentous truth. Here, dear reader, is the secret to being clutch: there is
no such thing as clutch performance ability. Dennis Eckersley
was simply a great pitcher *who didn't choke.*

Allow me to explain. In the eyes of the casual sports fan,
clutchness works something like this: with three seconds remaining in an NBA game, down by two points, Heroic Player receives
an entry pass behind the three-point line and begins to shoot.
Suitably charged with the gravity of the moment, Heroic Player
taps into a hidden recess of his soul and awakens the clutchness
beast within. Heroic Player now can transcend his typical 35 percent three-point shooting average and drastically enhance his
chances of hitting the shot; you just *know* he's going to rise to the
occasion and sink it. In this conception, then, clutchness is an
ability that mentally tough players can access in high-pressure
moments to push them above their normal ceiling.

Across all sports, however, there is zero solid evidence that
such an ability exists. Take clutch hitting in baseball. As a legion
of statistical whizzes have explained, if clutch hitting is truly an
ability athletes have, then it should persist from one year to the
next in the same way that players' batting averages tend to remain
steady between seasons. When statisticians look at a so-called
clutch hitter's performance in tight situations over time, however,
they see huge variations; great pressure performance one year
doesn't really predict pressure performance the following year.
Even the vaunted New York Yankee Derek Jeter, widely touted as
a clutchness paragon, has had his high-stakes hitting stats swing
wildly from season to season. None of this is to say that clutch
plays don't exist, or that some athletes aren't better under pressure

than others—clearly, they are. All this means is that no player has shown a statistically significant capability to *routinely exceed* his normal skill level under pressure.

This argument against clutch ability isn't the sort of thing you want to casually bring up with an athlete. When *Sports Illustrated*'s Tom Verducci tried to explain the clutch research to Jeter, he interrupted him midsentence. "You can take those stat guys," Jeter said, "and throw them out the window." But the figures don't lie: Jeter, it turns out, has virtually the same batting average in high-pressure situations as he has in low-pressure situations. Put another way, Jeter doesn't get *better* under higher stakes—he merely stays the same player. He's good early, and he's good late. He's just good. And indeed, when statisticians look at pressure performance figures, they find that the same players who are good the rest of the game (that is, the superstars) are the best in crunch time as well. In the NBA, for example, no one shined more brightly in the clutch during the 2007 season than Kobe Bryant, LeBron James, and Dirk Nowitzki—the three players who were the best in the league that year. A "clutch" player like Derek Jeter or Kobe Bryant, then, simply plays at his normal, extraordinarily high level of ability when the pressure is on. The moment doesn't unnerve him, and his talent naturally yields more big plays in clutch situations. "I know that under pressure, pitchers can get a little more velocity, players can run a little faster—those physical things," Rob Gray explained. "But you can't raise a skill above your normal ability. The really good clutch hitters just take every at-bat the same, whether it's spring training or the playoffs."

As performance researchers know, the case against clutchness leaves many people incredulous. How could anyone doubt clutch ability when we see players rise to the occasion all the time? Well, because what we *think* we see on the court isn't the whole truth—we're riddled with biases. In one influential 1985 study, for instance, three psychologists exploded the idea of the clutch-related "hot hand" in basketball. Nearly every athlete and fan (91 percent of them, to be exact) believes that basketball players can enter "the zone," a quasi-mystical "unconscious" state where shots flow with ease. Yet when these experts analyzed real basketball shooting data — Philadelphia 76ers field goals, Boston Celtics free throws — they found that the pattern of made shots didn't vary from chance. Just as a coin flipped a hundred times will sometimes come up heads six or seven times in a row, the players' near-even 52 percent shooting average meant that they'd occasionally hit several shots and appear "hot" when they were just adhering to statistical variation. The Harvard physicist Ed Purcell found the same thing in data on hot streaks in baseball: the streaks all fit with probability patterns according to the players' batting averages, with one notable exception to prove the rule — Joe DiMaggio's famed fifty-six-game hitting streak, in 1941. In judging clutchness, we also often fall prey to the "fundamental attribution error": attributing a person's behavior only to his character while disregarding circumstances and sheer luck, like the sudden gust of wind that barely lifted the game-winning homer out of the park.

But even though clutch ability probably doesn't exist as we typically conceive of it, "clutchness" is still a useful concept. After all, we know for sure that some players excel under pressure while others don't, and that many athletes who usually put up superstar numbers can become shockingly ineffective when the stakes mount. A player like Dennis Eckersley or Derek Jeter or Michael Jordan may not have a clutchness *ability*, but he still routinely outperforms the competition — even similarly talented players — in big moments. We don't need to toss out the entire concept of clutch; we just need to reassess what it means.

So what, exactly, *does* clutch mean? Sport psychologists who study high performers under pressure see a few common mental traits. First, they see self-confidence. Actually, to be honest, they see *over*confidence—a self-belief that exceeds their actual ability. Whenever researchers administer psychological surveys to athletes and then hunt for similarities among the highest achievers, they always find confidence to be the vital mental factor that distinguishes elite performance. We see this effect in other walks of life as well. For instance, the psychologist Gerald Matthews studies how people perform under stress in driving simulations, and his research consistently shows that cocky drivers perform best in tense situations: their arrogance muffles the effects of fear, similar to how Laurence Olivier's burst of ego upon seeing Noel Coward snapped him out of stage terror in the last chapter. "Confidence is kind of a protective illusion," Matthews told me. And why should an outsize bubble of self-confidence help? Athletes fail a lot, for one thing (nobody wins every time), so strong self-belief stops failure from getting them down. But beyond that, confidence also makes players feel they can succeed in a pressure-packed moment, which reduces the worry, self-doubt, and anxious internal monitoring that so often sabotage athletes. Self-confidence frees players to concentrate on their job and maximize their talent—thus, it helps them not choke.*

The second common theme among clutch performers flows from the first: if self-confidence makes athletes feel they can handle the pressure, they then see that ninth-inning at bat as a welcome *challenge* rather than a *threat*. Steve Sax, who played on two World Series–winning Dodgers teams, encapsulated this trait of clutch players succinctly. "They want the ball," he told me. "That's it. In the time of need, they want to be that person. I've seen other guys

---

*This confidence link may help explain why so many pro athletes are such unreformed egomaniacs, talking about themselves in the third person as though they're more force of nature than mere person: the profession self-selects for unrealistically high confidence.

who were afraid to fail and just wouldn't play in big games." Ronald Smith, a University of Washington psychologist who has done extensive research on high-stakes performance in baseball, agrees wholeheartedly. "Athletes who perceive a pressure situation as threatening and potentially disastrous tend to perform much more poorly than those who regard the same situation as a challenge or opportunity," he explained. "The ones who peak under pressure want the ball in the tight spot, whereas others pray they don't have to be in those situations because it's a tremendous ego threat." It's something we see in all clutch players, from Larry Bird to Tiger Woods: they covet the spotlight because they view a high-stakes moment as another welcome opportunity to shine.*

And now we can finally solve the mystery of a few pages back: how could the chronically nervous Dennis Eckersley—a man every bit as anxious on the mound as Steve Blass—possibly deliver such consistent clutch performances? Why should one terrified pitcher have excelled under pressure while another terrified pitcher melted down? After all, in so many ways, Blass and Eckersley were the same ballplayer; both were prodigiously talented, both had incredible control, and both felt tremendous performance anxiety. The only difference between them was how they *related* to that anxiety. Blass's fear sapped his confidence. As a Pirates pitching coach said of him in

---

*It's worth telling one more Larry Bird story here to illustrate the relationship between clutch performance and craving the spotlight. One game in Seattle, with only a few seconds left and the score tied, Bird stood in the huddle on the sidelines as his coach, K. C. Jones, drew up a play for fellow Celtic Dennis Johnson. But Bird had other ideas. "Why don't you just give me the ball and tell everyone else to get out of the way?" he interrupted. After the time-out, Bird approached the Seattle forward Xavier McDaniel and informed him, "Xavier, I'm getting the ball. I'm going to take two dribbles to the left. I'm going to step back behind the three-point line and stick it." Which is exactly what happened. Bird + confidence + a feeling of challenge and opportunity = Celtics win.

the midst of his struggles, "He has better stuff than most pitchers, but he doesn't believe it." Eckersley, on the other hand, always remained tenacious and cocky; even while anxious, he believed himself to be equal to any moment, and he almost always was. Blass responded to his fear by becoming avoidant, praying he wouldn't be called upon to pitch. But Eckersley was confrontational—he wanted the ball. After a subpar performance, he has said, "I couldn't wait to get out there to redeem myself." And then there was the choke factor. While Blass anxiously tweaked his throwing mechanics and let the internal focus disrupt his well-learned pitching motion, Eckersley did the opposite and treated every pitch the same, adhering to his routine and staying in the moment. "I step on the hill, and I sort of let it happen," he told the *San Francisco Chronicle*. Like all clutch athletes, Eckersley knew how *not* to choke.

And thus, the same fear produced two diametrically different results. One pitcher approached his anxiety with confidence, trust in his mechanics, and concentration only on the task at hand— and he became one of baseball's all-time-great closers. "My fear made me as good as you can be, the kind of fear that makes you stronger because you've got it channeled," Eckersley explained when he was inducted into the Baseball Hall of Fame in 2007. The other pitcher grew avoidant, insecure, and internally focused, and he became an emblem of choking.

But that's not the end of the story for Steve Blass.

Extra Innings

One day in the spring of 1999, Blass, by then a longtime Pirates broadcaster, was walking through the Pirates' spring training facility in Bradenton, Florida, when a peculiar man in cowboy boots approached him. This man, a New Mexico–based psychologist named Richard Crowley, had journeyed to Florida in hopes of finding and treating the Atlanta Braves pitcher Mark Wohlers, another sufferer of "Steve Blass disease." Instead, Crowley had stumbled on the man himself.

"Are you Steve Blass?" Crowley asked him.

"Yeah," Blass replied without breaking stride.

Crowley pursued him. "I've read about you," he hollered. "Do you still have that problem?"

"Yeah," Blass said, still walking.

"I could pop that," Crowley said.

Blass stopped and sized Crowley up. "I'd had tons of people approach me saying they could help," Blass recalled. "When he said he could fix it, I thought, *Who the hell do you think you are? Try living with this shit for a few years, and then tell me you can fix it.*" But Crowley persisted, hurriedly trying to explain his philosophy on treating performance anxiety. The skeptical Blass was intrigued just enough to grant Crowley a chunk of his time. They ended up talking for ninety minutes. After thinking it over for a couple of days, Blass agreed to meet Crowley again and go through a full therapy session.

Before going any further, I hasten to point out that Crowley's methods and psychological theories are more than a little bit loopy and offbeat. (As you may have guessed, it's not normal therapeutic practice for a psychologist to *seek out* a patient two thousand miles away from home.) His ideas wouldn't work for everyone. Yet Crowley's approach has helped several athletes, including major leaguers like Shawn Green and Wohlers (whom Crowley eventually tracked down). "These guys are already overloaded with information, so I don't go to the left brain with more information," Crowley told me. "Instead, I walk them into their right brain, particularly into their imagination." Crowley believes that Blass-style performance glitches are simple to fix. The player doesn't need to change his mechanics or probe his relationship with his mother, he says. The athlete just needs to find the negative thought that's dragging him down and remove it like a splinter. Crowley says he can do this in a single phone consultation—and this is where things get wacky. "I tell athletes to pretend they're like a kid and to bring out a mental image—a cartoon, a monster—that in their imagination is responsible for the ball going into the dirt," he said. "The image comes from the player. It might be a baseball bat with its tongue sticking out." Crowley has the athletes interact with that mental image until, finally, he tells them to destroy it. "They can use bazookas, pretend it's made of glass, make themselves a hundred feet tall and crush it," he told me. At this point, Crowley believes, the psychic poison has been extracted.

It's not the kind of scientifically proven treatment scheme that would pass muster in mainstream psychology, but Crowley's method inspired Blass. "Richard told me, 'You know what, we don't have to figure out *why* it happens, let's just fix it,'" Blass recalled. "Which was ideal in its simplicity. I thought, *Goddamn!*" After their session, Blass and Crowley walked down to the Pirates' field in Bradenton, and Blass told him that it was the first time he'd set foot there since spring training of 1975. "That was the field where he left his game," Crowley explained. Blass hadn't faced a batter even in the backyard since his departure from baseball in 1975, but that day he began the slow process of reclaiming the game he loved.

This journey started simply. Behind the Pirates' Florida clubhouse facility, far from any prying eyes, Blass tried playing catch with a broadcast colleague. Throwing a baseball, he suddenly remembered, was *fun*. Next, Blass asked a minor-league manager named Trent Jewett to meet him on the field early one morning. Jewett had once played catcher, and Blass wanted to toss some sliders and fastballs at him. So, for the first time in twenty-four years, Blass walked onto a pitching mound. He didn't need to worry about results or throwing mechanics; he just pitched because he felt like it. "I wanted to have just a little bit of that joy again," Blass said. "And Trent told me, 'Shit! You could pitch for me in Triple A!'" Soon, Crowley had the honor of being the first batter Blass threw against since 1975. The prospect still terrified Blass; the night before, he barely slept. In a batting cage at Bradenton the next day, Blass toed up and started hurling, anxious or not. "I did hit him a couple of times," Blass told me. "But the more I threw to him, the more I relaxed. I suddenly realized, *You know, this is not the end of the world.*"

The final step in Blass's reemergence came at the Pirates' fantasy camp that year. Chuck Tanner, a former Pirates manager, asked Blass to pitch in the annual game that the camp participants played against the former pros, but Blass resisted—until one lippy camper found his secret button. "One of the campers was a real pain in the ass," Blass recalled. "This guy told me, 'I could hit you right now. I think I could take you out of here.' So I asked Chuck to just let me pitch to this one guy and that would be it." Again, Blass felt apprehensive. Again, he couldn't sleep the night before the game. And then, for the first time in many years, the major-league pitcher Steve Blass showed up for a ball game. He tried to take everything slowly, in the moment, focusing on a cue word instead of his throwing motion. The first pitch to this smug camper felt fine. So did the second pitch. But the third pitch? Blass absolutely smoked it by him, putting the camper firmly in his place. *Okay*, Blass thought, *let's just stay out here.* He wound up pitching eight and two-thirds innings that day. "Every inning was more fun than the previous one," he told me. "I could have kept going and going. It was one of the finest days of my life. I didn't worry about how long it would last. I just went out, got the ball, and threw." Blass has pitched at the Pirates' fantasy camp each year since, soaking up every moment.

So Steve Blass was finally cured, thanks to Crowley's insistent help. After all, despite Crowley's somewhat wacky ideas, there *was* a method to his madness. Crowley's real gift to Blass was far greater than helping him to destroy some imaginary demon. (We've already seen the psychological forces that were bringing Blass down, and they certainly didn't include a cartoon bat with its tongue sticking out.) What he truly did was show Blass one powerful truth: in reality, there was never anything wrong with him. No mysterious "disease" afflicted him. He had no convoluted psychic problem to solve. By inspiring Blass to break free of his fretful mental clutter and take another shot at the game, Crowley helped unbind him from the chains of his anxious beliefs. All of the previous tweaking and figuring-out was unnecessary. Blass could sweep away the psychic mess, accept the anxiety he felt, and throw one pitch at a time. Simple.

This is the final lesson behind Blass's "disease." When he was struggling so mightily in his playing days, Blass was just a man who felt anxiety, like every one of us, and who followed that fear down a path to nowhere—doubting himself, avoiding uncomfortable situations, obsessing over his internal state, all of which merely drove him further astray. But with Crowley's help, Blass found a better place to focus his attention: on his love of the game. Once he resolved to approach his fear anew, mindfully and acceptingly, for the sake of reconnecting with the game he loved, baseball opened up for him again. From a point of deepest torment, throwing a ball could even become fun. And if anyone has ever earned that joy on the baseball diamond, it's Steve Blass.

CHAPTER SEVEN

Mayday, Mayday!: How We React, Think, and Survive When Our Lives Are on the Line

Hector Cafferata doesn't consider himself a hero for what he did. In fact, he resents the very implication—and since so many have lauded his extraordinary feat on that glacial Korean night six decades ago, the resentment has long since worn into exasperation. "Listen, none of that hero crap, okay?" he warned me, his voice inflected with his New Jersey upbringing, when I asked him to tell the story that eventually led to President Truman laying a Medal of Honor ribbon around his neck. "I'm just an average guy. If you make me out to be something I'm not, I'm not going to be a happy camper." It was a good-natured threat (at least I *hope* it was), but even in his dotage, Cafferata is the kind of man whose cautions prudent men heed. At age eighty, he's still built much like he was when he played semipro football in his youth: six foot two and two-hundred-plus pounds, with large, powerful hands and a stern no-bullshit stare. Cafferata has never been much for violence—"I've always been the kind of guy who don't want to hurt nobody," he told me—so when he reflects back on that night, the memory fills him not with pride at his valor under fire but with a feeling that he simply did his duty. "I was a marine, and marines are supposed to do their job," he said. "When the shit hits the fan, you take care of each other. That's it, period. And whatever you have to do, you have to do."

Heroic or not, what Cafferata "had to do" on the frigid evening of November 28, 1950, almost defies belief. At the time, Private Hector Cafferata belonged to a 240-man marine company that was stationed in the Chosin Reservoir, a mountainous and severe region in what is now North Korea. Just weeks before, the Korean War had heated up considerably when the Chinese entered the conflict to defend their fellow communists against advancing American forces—a development that caught U.S. commanders completely unprepared. With only 15,000 troops prowling through the Chosin Reservoir, the Americans suddenly had to contend with 150,000 Chinese soldiers looking to drive them out. That night, Cafferata's company won the unenviable task of defending a snowy corridor called Toktong Pass. As the sun drew low on the horizon, Cafferata's commander dispatched him and three other marines to a forward scouting position on a desolate hillside far from the American perimeter. Unbeknownst to anyone, a Chinese unit hid nearby, waiting to attack the pass under cover of night.

After arriving at their post, the marines split into two teams, with two of them taking the first watch while Cafferata and another New Jersey native, Kenneth Benson, dozed in their sleeping bags behind a nearby windbreak. An incredible, arctic-level cold surrounded them. Under the clear, moonlit sky, the temperature sank to 30 degrees below zero and an icy wind whipped over the snow. (The weather was so cold throughout the Chosin campaign that soldiers' canteens froze solid and medics had to keep morphine ampules in their mouths to prevent them from freezing.) The marines' youth and inexperience—Cafferata was only twenty-one—made them all a bit nervous, but the two soldiers on watch were especially jumpy that evening. On a couple of occasions, Cafferata awoke to hear them firing at shadows and tossing up flares, only to find nothing there. But the third time the gunfire broke out, at around one in the morning, Cafferata knew immediately that something was different. Soon, he heard screams and bugles and the clip of a machine gun. Bolting out of his sleeping bag, he barked, "Get up, Bens—they're here!" Just a dozen yards away from them, Chinese troops flooded down the hillside. Cafferata put the butt of his M1 rifle to his shoulder and began firing.

The fight was nearly over before it even began. Within moments, the Chinese incapacitated the two marines to Cafferata's left. Meanwhile, he and Benson quickly realized that they were in a terrible defensive position, so they rushed back toward a wash that provided better cover. There, they stumbled upon several injured marines taking refuge. "These boys were badly wounded," Cafferata recalled. "They got hit real quick, before they could wake up. We couldn't leave 'em, so we decided to make our stand there and protect them." But soon, that "we" dwindled to an "I." When a Chinese grenade landed in their midst and Benson tried to hurl it back, the grenade exploded as it left his hand, blowing his glasses off his face. The blast left Benson bloodied and severely flash-blinded; Cafferata was going to have to hold off the Chinese unit on his own. Under such crushing odds, many soldiers would have raised the white flag here, but not Cafferata: "I always felt that if you wanted my ass, you better bring your lunch," he said.

For the next seven hours, Cafferata became a one-man fighting force. With moonlight and flares providing illumination, he hustled up and down the wash, taking out advancing Chinese troops. In his left hand, Cafferata carried an entrenching tool, which he used to bat away any Chinese grenades that flew in. "I was the world's worst baseball player, so I don't know how I hit them," he told me. He held his eight-shot M1 Garand in his right hand, laying it across his left thumb to shoot, picking off the ill-trained Chinese soldiers one by one. Whenever Cafferata ran out of ammo, he grabbed another rifle from Benson, who was reloading them by feel. Cafferata was scared, of course—or, as he phrased it, "I was real concerned, put it that way"—but he doesn't remember contemplating his feelings too much that night. (Although he does recall thinking that his mother "was going to be really pissed off" if he got himself killed.) Instead, with Chinese bullets and grenades hailing in, he acted on intuition. "I tell you, you don't have much time to think," he said. "You react automatically. I guess you're operating on instinct, or how you grew up." Ever since Cafferata was ten years old, he had hunted and target-shot with rifles more or less constantly; now, he found, all he had to do was level his gun and pull the trigger—it all came so naturally. "Shooting to me was like eating," he told me. "I grew up with a gun in my hands. So my shooting wasn't luck."

Cafferata fired his rifle so much that night that the barrel began to blacken and catch fire; he had to cool it down with snow. Late in the battle, he tried to fling a grenade back manually but it blew too soon, badly damaging his nearly frozen left hand. He kept fighting. Sometime after dawn, marine reinforcements arrived to find a single man holding off an entire enemy unit, as if possessed by supernatural energy. It wasn't until this point, when he could relax a little, that Cafferata discovered that he'd fought all night in his socks, and without his parka. Suddenly realizing that his feet were like two hunks of ice, he went looking for his boots and, finally, got hit by enemy fire in his chest and his right arm. After Cafferata's fellow marines evacuated the injured and collected dumbfounded Chinese prisoners, they set about tallying the enemy dead. The results stunned them. As one soldier explained to the writer Larry Smith many years later, "I figured he killed close to a hundred of the enemy but we only wrote it up for thirty-six because we didn't think [anyone] would believe it."

Naturally, from a loss-of-human-life standpoint, this story of a hundred soldiers dying on a frigid mountainside is tragic—a fact that Cafferata is the first to point out. "I felt sorry for them," he told me. "I didn't enjoy shooting 'em, but what could I do? If I didn't shoot them, they'd shoot me." From a performance-under-fire standpoint, though...I mean, *good God*. What Cafferata pulled off that night, with mortal peril hanging constantly overhead, was such an incredible feat that it seems like something out of a video game: all alone, outnumbered a hundred-plus to one, he held the American perimeter. How could one terrified human being, acting on pure instinct, function so well for so long against such staggering odds? Sixty years have now passed since that night, and Cafferata has had plenty of time to reflect on his deed. He spent eighteen months convalescing in military hospitals from his injuries, which left his right hand nearly useless (though he can still pull a trigger), and he reluctantly accepted a Medal of Honor from President Truman for his actions. (As he likes to tell it, "the little bastard" ruined his dress shoes by stepping on them in his attempts to reach around Cafferata's neck.) But still, after all these years, Cafferata finds himself at a loss to explain his incredible conversion into a one-man army when he was in the most extreme danger imaginable. As he once put it, "To tell you the truth, I did it. I know I did it. Other people know I did it. But I'll be goddamned if I know how I did it."

So far in this book, we've mostly explored fear as an emotion with an evolutionary purpose that is almost comically mismatched with the requirements of modern life: it's designed to help us deal with threats to life and limb, yet none of us is going to *die* if we deliver an anxiety-provoking speech or drop a last-second touchdown pass. From this perspective, fear can seem like some sort of mental vestige from the days when mastodons and saber-toothed tigers menaced us on a daily basis. Yet as sheltered as we are today, sometimes our extreme, self-protecting fear reactions truly do match the situations we're in. Sometimes a grave, life-threatening crisis springs suddenly into your path and you really *are* going to die if you don't act swiftly and wisely. How, then, do we respond in these extremes of fear, with our survival hanging in the balance? As Hector Cafferata found out that frozen November night, when the proverbial shit hits the proverbial fan, we turn into utterly different people.

"You'll Be Scared"

In September 1895, Stephen Crane's classic Civil War novel *The Red Badge of Courage* thundered onto the American literary scene. As the public snapped up copies, critics acclaimed Crane's book for its realistic portrayals of soldiers' harrowing battlefield experiences. The novel tracks the introspective Union soldier Henry Fleming's evolution from cowardly deserter into brave fighter, and to virtually everyone, Crane's depiction seemed so accurate a portrait of war that it may as well have been nonfiction — but not to one grizzled Union veteran named Abner Small. Crane was an imaginative writer, Small pointed out, but he'd never actually seen combat; in fact, the twenty-three-year-old Crane wasn't even born until six years after the war ended. One aspect of the novel that particularly struck Small as "sheer rot" was that Crane seemed to believe Fleming could *think sensibly* in the midst of war. "A young author, his boyish fancy stimulated by books, has written a story in which he takes a raw recruit into, through, and out of a battle and represents him with a brain fully alive to reason," the peeved Small wrote in his war memoir. "That any man in my regiment, or in the army, analyzed his feelings and marked out any specific line of conduct while under fire, or even thought for five consecutive minutes of the past, present, or future...is absurd." Small believed that only those who have endured the terror and chaos of battle could understand the truth: when you're in that much danger, the standard rules of human psychology fall apart. No one is "normal" under fire.

It took another fifty years before the rest of the world began to catch up with Small's insight. For decades, old-line military commanders clung to archaic, idealized beliefs about how soldiers should behave on the battlefield. If properly prodded, all men were capable of cool-headed boldness with bullets whizzing by, they declared—it was just a matter of instilling discipline, a sense of duty, and dedication to country. Fear was weakness, shameful, never to be seen or spoken of; in World War II, for instance, all gun sites on Malta bore a posted message warning troops, "If you are a man you will not permit your self-respect to admit an anxiety neurosis or to show fear." Deserters and stragglers were beneath contempt, and they deserved the harshest punishment possible. (After World War II, writes Joanna Bourke in *Fear: A Cultural History*, some even argued that those who had panicked in battle should be forcibly sterilized, to prevent their weakness from passing to future generations.) In the British military, officers long carried pistols not to fire upon the enemy but to shoot those who refused to advance into enemy fire.

But as times grew more enlightened, psychology researchers began revealing just how twisted the old-fashioned views of fear in combat really were. The watershed moment came in the Second World War, when several studies shattered outdated assumptions about battlefield bravery. Regardless of rank, experience, and level of daring, they found that *everyone* gets terrified in battle. According to one series of interviews with World War II infantrymen, only 7 percent of the men said they never felt afraid—yet 75 percent of the men admitted that their hands trembled in battle, 85 percent got sweaty palms, and 89 percent had pre-combat insomnia. Perhaps most revealing of all were the surveys in which an eighth of all U.S. soldiers said that they had lost bowel control in battle, while a quarter had lost bladder control.* Soon, even General George Patton, by no means a warm-and-fuzzy mollycoddler (he once lost his command because he smacked a soldier who was in the hospital for shell shock), felt comfortable speaking frankly about the topic. "Every man is scared in his first action," he said. "If he says he's not, he's a goddamn liar....The real hero is the man who fights even though he's scared." The U.S. military was particularly candid about fear in World War II, and many have speculated that this was why fewer American vets developed PTSD than their stiff-upper-lipped British counterparts. One handbook, given to all recruits, blared in capital letters, "YOU'LL BE SCARED," and continued like an overprotective mother, "Don't let anyone tell you you're a coward if you admit to being scared."

In many ways, these findings were far from earth-shattering; who could really be surprised to learn that war was horrifying? But in the same time period, two inventive researchers—not academics, but soldiers—performed field studies of combat troops that shook up the established ideas about how the human mind responds to extreme danger. The more notorious of these two men was the combat historian S. L. A. Marshall, a World War I army grunt and twenty-year newspaperman whom the U.S. Army tasked with gathering oral history data from soldiers fighting in World War II. Although military scholars have often questioned the rigor of Marshall's research methods (especially after

---

*As one veteran told the combat expert Colonel Dave Grossman—whom we'll meet later in this chapter—upon hearing this factoid, "Hell Colonel, all that proves is that three out of four are damned liars!"

he made the hugely controversial assertion that 75 percent of American troops never fired their weapons at the enemy during WWII because of their innate revulsion over killing), his work has massively influenced the way the military views soldier psychology—particularly in regard to fear.

For Marshall, the battlefield was a place of utter chaos, a maelstrom of danger and disorder that made soldiers feel an unpredictable range of emotions. Engagements with the enemy seldom lent themselves to cut-and-dried psychological analysis. "The same group of soldiers may act like lions and then like scared hares within the passage of a few minutes," he observed in his classic book *Men Against Fire*. To illustrate this, Marshall tells the story of one battle-weary U.S. Army patrol unit on night duty in the Normandy countryside. At the base of a hill, this unit stumbled onto a German machine-gun post, sending the men into panicked flight; they "ran like dogs," Marshall reports, and two commanding officers had to "beat them back with physical violence" to prevent them from fleeing entirely. One officer, Lieutenant Woodrow Millsaps, spent a full hour commanding the terrified soldiers to do their duty, then literally begging them to attack. When this petrified unit finally advanced on the Germans, Marshall writes, the results were startling:

At last they charged the enemy, closing within hand-grappling distance. The slaughter began with grenade, bayonet, and bullet. Some of the patrol were killed and some wounded. But all now acted as if oblivious to danger. The slaughter once started could not be stopped. Millsaps tried to regain control but his men paid no heed. Having slaughtered every German in sight, they ran on into the barns of the French farmhouses where they killed the hogs, cows, and sheep. The orgy ended when the last beast was dead.

Instantaneously, these men swung from fiercest terror to deepest rage.* Marshall concedes that this incident was unusual, but adds, "In battle the unusual is met usually and the abnormal becomes the normal."

Marshall's investigations into the psychological mysteries of the battlefield also shed light on another side of the mind's response to extreme danger: the importance of human contact. Once, an army general asked Marshall to solve a battlefield riddle. Why was it, the general wondered, that when a row of advancing infantry suddenly hit enemy fire, the troops—even the hardened veterans—often stopped cold for up to an hour before resuming their forward push? After Marshall quizzed soldiers from eleven infantry companies, a single root cause soon emerged. The problem arose, he found, when the troops couldn't see each other— they felt frighteningly alone, and this temporarily paralyzed them. Losing visual contact with buddies under fire produces "moral disintegration" of the line. Wrote Marshall, "Be a man ever so accustomed to fire, experiencing it when he is alone and unobserved produces shock that is indescribable."

It's a finding that extends to all crisis behavior: in an emergency, we invariably seek out the company of others to reduce our stress and fear. Immersion in a group makes us feel safer (as long as we're not performing in front of it, that is), and the better we know our companions, the greater the comfort we derive. In one 2006 study, for instance, the psychologist Richard Davidson

---

*Because the fight and flight impulses spring from the same source, the amygdala, our emotions travel the fear-anger superhighway with surprising speed. This is why military leaders have historically tried to send troops into a state of fury with prebattle speeches: anger often counteracts fear. It's also why Laurence Olivier used to soothe his stage fright by peeking out at the audience from behind the curtain and softly cursing, "You bastards... you bastards..."

performed a brain imaging study on sixteen married couples in which he placed the woman under threat of electric shock either (a) when alone, (b) when holding a stranger's hand, or (c) when holding her spouse's hand. Holding the stranger's hand reduced each woman's anxiety level compared to being alone, but holding her husband's hand pushed her feelings of fear and discomfort down significantly—and the amount it helped corresponded directly with the quality of their marriage.

In the Second World War, the social relationships between soldiers were so momentous that they overpowered nearly every other emotion. There was only one fate most soldiers feared worse than injury and death, Marshall found, and that was the thought of showing cowardice in front of their comrades; not even fighting for a cause could inspire the same level of courage as personal honor. Fear proves no match for the social bonds between us. Indeed, we're often willing to die to preserve them.

One dusty day in Iraq's Diyala province, the U.S. Army specialist Alex Horton discovered this fact firsthand. In May 2007, Horton was serving with the Third Stryker Brigade, stationed near the insurgent-controlled city of Baqubah. Every day, his unit patrolled the Baqubah streets, hoping to draw enemy fire. "We were basically live bait," Horton explained to me. "Sometimes, they would tell us to go out and not come back until we were shot at." For the eleven months of Horton's tour to that point, in Baghdad and Mosul, he'd done these patrols from the relative safety of a Stryker, an armored, eight-wheeled combat vehicle that looks like a trimmed-down tank. After he arrived in Baqubah and his platoon discovered fifty improvised explosive devices (IEDs) littering a single one-mile stretch of road, however, they had to start patrolling on foot. That May, Horton's unit was walking across a vast, empty field when the rapid crack of a machine gun burst out nearby. Bullets hissing by their heads, Horton and his comrades sprinted toward an enclosed courtyard fifty yards away. Just after Horton vaulted through the courtyard gate, another soldier lunged through screaming *"Fuck! Fuck! Fuck!"* as machine-gun bullets punched holes in the concrete right above his head. "It was like something from a cartoon," Horton told me. "The bullets made an outline of his head in the wall." His team leader peered out to try to get a fix on the gunner's position, and bullets immediately clipped by his ear. They tried another exit but met fire from a second gunner there. Said Horton, "We were pinned. There was no way of getting out of it without someone getting shot."

The platoon radioed to ask for helicopter support, but the request got rejected; the higher-ups decided that it would scare the insurgents off, and they wanted them dead or alive. Instead, a commander ordered Horton's platoon to charge the machine guns—a move that seemed like certain death. At that moment, Horton learned just what the fear of appearing cowardly could drive a soldier to do. "Even though I thought it was suicide, that it was crazy, that there was no way we could pull it off, I was prepared to run out there," he recalled. "I mean, you have a fear of getting mowed down by a machine gun, but there's a greater fear of looking like a pussy to your friends. If it's a choice between looking unreliable to everyone and eating a bullet, then eating the bullet is the most viable option." As they were preparing to sprint out, though, the orders changed. The platoon lobbed smoke grenades into the field and dashed to the side, ducking fire and leapfrogging from house to house until they reached a flanking position. Before they could draw close, the insurgents bolted. Improbably, Horton and his fellow troops made it through unscathed, and to this day, he remains shocked at how willing he was to sprint needlessly into fire for the sake of his social ties— just as Marshall would have predicted.

Wolves and Sheep

While Marshall studied the deeds of American troops in Europe, the era's other innovative combat investigator found himself at the southern edge of that continent, working toward a realization that would become a bedrock principle of crisis psychology. In July 1943, Lieutenant Colonel Lionel Wigram, the leading light of the British Battle School, traveled to Sicily to study the performance of Eighth Army infantry units as they attacked the Italian front. After observing countless assaults, Wigram began to pick up on a peculiar pattern in each twenty-two-man platoon. Whenever a platoon ran into enemy fire, Wigram noted, the troops would react in three very distinct ways. Without fail, a few soldiers would go to pieces and "start making tracks for home." Another handful of "gutful" men would respond valiantly, opening fire and advancing. And the majority of the troops? Wigram said they would enter a state of bewilderment, unsure of how to act; they became "sheep," and would only do their duty when prodded by a strong, decisive leader. No matter the unit, Wigram saw the troops gravitate toward these three responses in exactly the same proportions. "Every platoon can be analysed as follows," he wrote in his report on the phenomenon. "Six gutful men who will go anywhere and do anything, twelve 'sheep' who will follow a short distance behind if they are well led, and from four to six ineffective men who have not got what it takes in them ever to be really effective soldiers."

To say that Wigram's discovery was provocative would be a titanic understatement. Essentially, Wigram was asking the British military establishment to completely revise its expectations of its fighting forces. Instead of hewing to the mistaken belief that all men are capable of gallant heroism with their lives on the line, he argued that the army should just accept that every platoon inevitably contained a certain percentage of cowards and sheep and plan accordingly. Wigram even proposed a two-pronged strategy for minimizing troop terror under fire: shift the panickers off the front lines into support jobs and find ways to bind the sheep to the heroes so they'd be more inclined to follow their lead. But Wigram's idea proved too revolutionary for the British military brass's taste, especially in the middle of a world war. Believing his report would hurt troop morale, General Bernard Montgomery buried it and demoted Wigram, shuttling him off to lead an army company in Italy, where he later died leading an assault.

Wigram's insight into the behavior patterns of platoons under fire provided an astonishingly accurate model for how *all* of us react to a sudden emergency. In fact, the British psychologist John Leach has spent several decades researching human crisis behavior in scenarios ranging from natural disasters to fires to airplane crashes, and his studies have shown that groups react to a life-threatening disaster in virtually the same proportions Wigram outlined. When calamity strikes, Leach says, we transform from our normal selves into our crisis selves. Ten to twenty percent of us will remain cool and composed, he writes in his book *Survival Psychology:* "These people will be able to collect their thoughts quickly, their awareness of the situation will be intact and their judgment and reasoning abilities will not be impaired to any significant extent." Another 10 to 15 percent will completely freak out, reacting in ways that actually endanger their survival—"uncontrolled weeping, confusion, screaming and paralyzing anxiety."

But the vast majority of us—70 to 80 percent, by Leach's estimates—become the bewildered sheep. In the direst crisis, our mental circuitry often overloads, leaving us stunned and uncertain of what to do next. The most common behavior in response to a terrifying emergency isn't decisive action or panicked screaming,

Leach says.* No, what you generally see from people immersed in a crisis is this: dazed, paradoxical lethargy. Take the reaction of workers in the Twin Towers on 9/11, for instance: a series of interviews with almost nine hundred World Trade Center survivors revealed that people waited an average of six minutes after the planes hit before beginning to descend the stairs to safety. While a few quick thinkers bolted immediately, others dawdled for up to thirty minutes; they took the time to shut down computers and gather belongings, clinging to their normal routines. And when they finally *did* get to the stairs, these survivors took a full minute to go down each floor—double what emergency planners had predicted. They were neither panicked nor calm, but simply stunned; as one woman put it, "We were like robots."

Blindsided by a horrifying crisis they'd never foreseen, these World Trade Center workers greeted extreme danger in a very human way: they couldn't believe it was really happening. Fueled by what psychologists have deemed "normalcy bias," humans in crisis have a troubling tendency to deny that anything out of the ordinary is going on. With scientists proclaiming that a mammoth typhoon or volcanic eruption looms close at hand, residents refuse to evacuate; just look at the thousands who stayed behind in New Orleans before Hurricane Katrina, or the hangers-on near Washington's Mount Saint Helens as it prepared to blow sky-high in 1980. Leach witnessed such denial at work in November 1987, when he was on the scene for the deadly London Underground fire at King's Cross station. As Leach tried to

---

*By all accounts, the mass hysteria one sees in Hollywood movies—herds of terrified citizens barreling wildly down the streets while clawing at one another—almost never happens. The disaster researcher E. L. Quarantelli, who has performed more than seven hundred field studies, goes so far as to call widespread panic a media-created myth that has no place in social science. It turns out that people are mostly quite polite in disasters.

warn commuters about the smoke pouring visibly out of the tunnels, people blithely followed their routines and shoved their way into the burning complex—even though only yards away, some were literally engulfed in flames. Baffled and poorly trained Underground staffers didn't even spray a single fire extinguisher as trains continued arriving and disgorging passengers directly into the blaze.*

This sheeplike behavior does differ from the freezing reaction we learned about in Chapter One (sheep may move slowly, but at least they *move*), but disaster researchers see plenty of maladaptive freezing in emergencies as well. Far too often, people confronted with a crisis that requires urgent action do nothing to save themselves, as occurred in the legendary Beverly Hills Supper Club inferno, which swept through a Cincinnati-area nightclub in May 1977, killing 167 people. Even as dense smoke curled into the club's dining rooms and staff members shouted warnings, patrons were astonishingly slow to flee. Sifting through the wreckage after the fire died down, rescue workers discovered six charred bodies perched in their chairs around a table; they'd died just sitting there. Or consider the famed March 27, 1977, crash at the airport in Tenerife, Spain—a historically catastrophic aviation disaster with 583 fatalities, yet also one in which many people died unnecessarily. That afternoon, a KLM-operated 747 landing amid thick fog sliced into a Pan Am jet waiting on the runway to take off. Those on the KLM aircraft all died instantly, but many on the clipped Pan Am plane survived the initial impact and had at least a minute to flee before their jet completely filled with flames. The few who scrambled off the plane to safety later reported that their fellow passengers simply

---

*Because of lackluster responses like this, today many safety agencies train workers such as flight attendants and train employees to shout authoritative instructions at people in an emergency, to break them out of their stupor and get them moving.

sat there frozen, wide-eyed, mouths literally agape, as the inferno crept around them. The accident, sudden and catastrophic as it was, seemed to be too much for their minds to process quickly.

This treacherous freezing reaction is startlingly common in crises. In the 1970s, the psychologist Daniel Johnson found that when he asked subjects to perform a challenging and novel task under high pressure, 45 percent of them shut down and stopped moving for a minimum of thirty seconds. "They quit functioning," he later told the *Time* reporter Amanda Ripley. "They just sat there." Under life-threatening stress, complex brain processing plummets and the neural mainframe often maxes out, leaving us with instinctive reactions and deeply worn routines; it's as if such an experience can be too monumental for the brain to compute. Leach says that even in the most ideal of circumstances, the brain needs at least eight to ten seconds to process an unfamiliar situation—but intense fear and stress make that task even tougher. When our brains search their data banks for the right course of action in extreme danger and come up empty, Leach claims, they shut down. We freeze. We do nothing.

Maladaptive as they are, these reactions to mortal peril—from ill-advised freezing to livestock-esque bewilderment—don't reflect any large-scale failings in personal character. They merely reflect the way our minds are programmed. When gravest disaster catches us unprepared, our normal faculties depart and we really do become different people: impulsive, overloaded automatons, for the most part. The person we *think* we are may not be the person who appears under fire. Even our experience of reality itself might change.

The Distortions

In the mid-1990s, the clinical psychologist Alexis Artwohl began counseling officers in the Portland, Oregon, police department after they had been involved in shootings. Unlike most psychologists, Artwohl had a special rapport with these cops because her husband was an assistant chief in the department, and her empathy for their lives soon led them to divulge things—very *odd* things—that they usually hid from therapists, other officers, and even their families. Within Artwohl's office, the cops spoke honestly about peculiar sensations they had experienced in crises, like seeing the world as if through a keyhole, or feeling as if they were standing outside their own bodies. "These officers had these strange experiences in stressful situations but didn't feel like they could talk about them openly," Artwohl told me. "They thought there might be something wrong with them, or they were the only ones to whom this had happened. And of course we now know that's not true at all—these perceptual distortions are quite normal."

Over a period of five years, Artwohl gave hundreds of officers a written survey to fill out about their shooting experiences. Her findings were remarkable: Virtually all of the officers reported experiencing at least one major perceptual distortion. Most experienced several. (A similar study by the criminologist David Klinger put the number of officers who experience a minimum of one form of altered reality under fire at 94 percent, with 89 percent reporting more than one.) For some, time moved in slow motion. For others, it sped up. Sounds intensified or disappeared altogether. Actions seemed to happen without conscious control. The mind played tricks. One officer vividly remembered seeing his partner "go down in a spray of blood," only to find him unharmed a moment later. Another believed a suspect had shot at him "from down a long dark hallway about forty feet long"; revisiting the scene a day later, he found to his surprise that the suspect "had actually been only about five feet in front of [him] in an open room." Wrote one cop in a particularly strange anecdote, "During a violent shoot-out I looked over…and was puzzled to see beer cans slowly floating through the air past my face. What was even more puzzling was that they had the word *Federal* printed on the bottom. They turned out to be the shell casings ejected by the officer who was firing next to me." None of these perceptual warps had ever been researched in any systematic way before. "With all the years of police action and combat, it's surprising to me that this wasn't more widely known." Artwohl shrugged.

The single distortion under fire that Artwohl heard about most, with a full 84 percent of the officers reporting it, was diminished hearing. In the jarring, electrifying heat of a deadly force encounter, Artwohl says, the brain focuses so intently on the immediate threat that all senses but vision often fade away. "It's not uncommon for an officer to have his partner right next to him cranking off rounds from a shotgun and he has no idea he was even there," she said. Some officers Artwohl interviewed recalled being puzzled during a shooting to hear their pistols making a tiny pop like a cap gun; one said he wouldn't even have known the gun was firing if not for the recoil. This finding is in line with what neuroscientists have long known about how the brain registers sensory data, Artwohl explains. "The brain can't pay attention to all of its sensory inputs all the time," she said. "So in these shootings, the sound is coming into the brain, but the brain is filtering it out and ignoring it. And when the brain does that, to you it's like it never happened."

The brain's tendency to steer its resources into visually zeroing in on the threat also explains the second most common perceptual distortion under fire. Tunnel vision, reported by 79 percent of Artwohl's officers, occurs when the mind locks onto a target or threat to the exclusion of all peripheral information. (Leach, the survival psychologist, says some evidence suggests that tunnel vision also happens in another situation that ensures humankind's continued existence: during sex.) Studies show that tunnel vision can reduce a person's visual field by as much as 70 percent, an experience that officers liken to looking through a toilet paper tube. The effect is so pronounced that some police departments now train their officers to quickly sidestep when facing an assailant, on the theory that they just might disappear from the criminal's field of sight for one precious moment.

According to Artwohl's findings, the warping of reality under extreme stress often ventures into even weirder territory. For 62 percent of the officers she surveyed, time seemed to lurch into slow motion during their life-threatening encounter—a perceptual oddity frequently echoed in victims' accounts of emergencies like car crashes. In a 2006 study, however, the Baylor University neuroscientist David Eagleman tested this phenomenon by asking volunteers to try to read a rapidly flashing number on a watch while falling backwards into a net from atop a 150-foot-tall tower, a task that is terrifying just to *read* about. This digit blinked on and off too quickly for the human eye to spot it under normal conditions, so Eagleman figured that if extreme fear truly does slow down our experience of time, his plummeting subjects should be able to read it. They couldn't. The truth, psychologists believe, is that it's really our *memory* of the event that unfolds at the pace of molasses; during an intensely fear-provoking experience, the amygdala etches such a robustly detailed representation into the mind that in retrospect it seems that everything transpired slowly. Memories, after all, are notoriously unreliable, especially after an emergency. Sometimes they're eerily intricate, and yet other times vital details disappear altogether. "Officers who were at an incident have pulled their weapon, fired it, and reholstered it, and later had absolutely no memory of doing it," Artwohl told me. If your attention is focused like a laser on a threat (say, the guy shooting at you), Artwohl says, you may perform an action (such as firing your gun) so unconsciously and automatically that it fails to register in your memory banks.

Perhaps the strangest perceptual distortions of all, however, are the ones that can make a previously calm, clearheaded person suddenly detach from reality—and for the most striking demonstration of this, we turn not to people under fire, but to people under*water*. If you peruse the reports on fatalities in amateur diving (which, no doubt, is one of your favorite pastimes), a strange statistic might pop out. Investigators chalk up 60 percent of underwater deaths to the causes we'd expect, like faulty breathing equipment or health problems, yet 40 percent are listed as "unexplained"; these divers somehow drowned even though they had suffered no major injuries and their tanks still held plenty of air. When the psychologist William Morgan began probing this data, he noticed something still more puzzling. Some divers had torn out their oxygen regulators for no apparent reason at all, gasping in water as though it were clean air. Others even attempted to rip out the regulators of their horrified fellow divers. Why on earth would they do such a thing? Frightened or highly stressed divers, Morgan found, occasionally experienced a deep instinctive feeling that they were *suffocating* when their regulators covered their mouths and noses, and in their blind panic they automatically did what they'd do on land: remove the obstruction from their airway. Terror drastically distorted their interpretation of reality, triggering an instinctive response that was fatally mismatched to their true situation.

Reacting to extreme danger without conscious thought often saves our lives, but these divers' lethal distortions also show how our instincts can make a hazardous situation far worse. Bernie Chowdhury, a veteran technical diver, has trained aspiring aquanauts for twenty years, and before his students can venture safely into the undersea wonderland of submerged cave systems and loot-filled shipwrecks, he must drill out their dangerous land-dweller instincts. (We didn't evolve to spend much time underwater, of course, just as we didn't evolve to get in gunfights or escape burning skyscrapers.) "We have many, many eyewitness accounts of people doing things underwater that are counterproductive to their survival, and then they die," Chowdhury told me. "The fight-or-flight response is not a good thing underwater. In technical diving, it can cripple you or seriously injure you, if not kill you." For example, say a diver runs into a problem with his breathing apparatus a hundred feet down and begins to feel the danger of suffocation closing in. Here, his impulse will be to make for the surface, stat. But when you're breathing a compressed air mix from a tank, you can't do this; the gas builds up in the body's tissues, and unless you ascend and decompress slowly, sometimes over the course of several *hours*, the gas will expand under reduced pressure and cause the dreaded ailment divers call "the bends." To survive, technical divers train themselves to cut through perceptual distortions and go against ill-advised instinct, working their dilemmas out in the deep. Said Chowdhury, "Underwater problems have to be solved underwater."

In fact, psychologists say that even the most paltry shred of preparation can drastically increase our likelihood of responding adaptively in a crisis. Many of those who escaped that blazing Pan Am plane in Tenerife were able to break out of their shock and act fast simply because they studied the flimsy safety information card nestled in the pouch in front of their seat. By loading the exit locations into their brains before disaster hit, they were able to cut their reaction time from the eight-plus seconds that unprepared people require to process a novel situation down to a mere one hundred milliseconds; with nothing for their minds to figure out under fire, they could move out immediately. We may not be able to reason like champions when disaster strikes, and we might find our experience of reality unhelpfully distorted, but we do have one means of stacking the odds in our favor. Through training, we can preload our instincts.

An Ounce of Prevention...

The combat expert Dave Grossman likes to tell a story to illustrate the vital role that training plays in performance under fire. Grossman, a former army ranger and psychology instructor at West Point, occupies a privileged and somewhat unusual position in the study of extreme human performance: as the nation's preeminent lecturer on battlefield behavior, he's on the road three hundred days a year speaking with soldiers and cops who have brushed up against mortality. As Grossman told me with more than a glimmer of pride, "Every day, I get to talk to someone who's been there. Every day. And I believe I've talked to more human beings who have had to kill in combat than anybody alive— maybe anybody in history." On one of these stopovers, a police officer told him about a shooting experience that, for Grossman, perfectly encapsulated how powerful preparation can be when a crisis suddenly strikes. On a routine patrol, the officer recounted, he spotted an improperly parked van and walked over to ask its driver to move. "I didn't know that he'd already killed one person," the officer later said. The man quickly drew his gun, recalled the cop, "Then a hole appears in his chest and the guy drops. My first thought was, *Whoa, somebody shot him for me!* I actually looked over my shoulder to see who shot this guy. Then I realized I had my gun in my hand and it was me who had shot him." The cop acted on instinct, built up through practice and time on the firing range, and it saved his life.

This is the kind of story that fires Grossman up, because it so flawlessly exemplifies the core lesson he strives to teach: "In combat, you do not rise to the occasion—you sink to the level of your training." None of us can safely rely upon fantasies of our innate heroism when the moment calls for it, Grossman says, because a life-threatening crisis makes our normal experience of reality fall apart. Only the man who prepares for disaster can hope to respond well. "Do not expect the combat fairy to come bonk you with the combat wand and suddenly make you capable of doing things that you never rehearsed before," Grossman warns in his book *On Combat: The Psychology and Physiology of Deadly Conflict in War and in Peace*, co-written with Loren Christensen. Grossman often points out that humans are a unique hybrid of predator and prey. "We've got the neural network of prey, to run like hell and get out of there," he told me. "But we also have the gripping fangs and forward-set eyes of a predator, so we have that neural network as well. Now, at the moment of truth, which one are you going to be? When you're in a gunfight, are you the predator or the prey? Well, it all depends on which set of neurons you've exercised." If you find yourself needing to figure out for the first time what to do when your life is on the line, chances are you'll become the prey. Survival under fire is about formatting your brain to take the right action reflexively—that is, it's about only sinking to the level of your training, and no lower.

From cops in mock shoot-outs to ordinary people stuck in fire drills, training must be as realistic as possible for it to be effective. For decades, all the way up through both world wars, the United States military prepared its soldiers for battle by having them practice shooting at circular targets, and this, Grossman believes, accounts for Marshall's famous claim that 75 percent of American troops never fired on the enemy. Quips Grossman in *On Combat*, "The fundamental flaw in training for combat in this way is that there are no known instances of any bull's-eye targets ever attacking our warriors." In other words, soldiers who have never rehearsed shooting at something resembling a human being can't be relied on to summon the will to do something so unfamiliar and unnatural in true warfare.

And indeed, after Marshall dropped his bombshell, the U.S. military adapted its training methods. As Alex Horton, the army soldier who got pinned in the Baqubah courtyard, recalled of his instruction at Washington's Fort Lewis, "Some of our training was just getting used to the idea of shooting someone." Today, drill sergeants run recruits through endless playacted scenarios before they ever see a single second of live action. "The training we did was as realistic as it could possibly be," Horton explained. "They'd have actors play bad guys and civilians, and we'd set up scenarios where we had to find some guy and ask for information, kind of a game. Then they'd throw a training grenade out and say, 'Okay, you got hit by an IED, two wheels are off, two of your guys are seriously injured,' and we'd have to respond as if it were really happening." Cops get the same training treatment nowadays, through live-action exercises in which they fire "simunition" at playacted criminals. These sessions don't completely approximate combat, Horton says, but they're close. "You don't feel like you'll be ready for combat until it actually happens, but when it does, it feels familiar," he told me. Which is why today's U.S. military, far more well prepared than any fighting force in history, shows fewer maladaptive reactions in combat than the hastily and archaically trained troops Wigram and Marshall studied.

Because trainees respond to a crisis so similarly to the way they've rehearsed, they must be very careful of how they conduct the small details of their practice, as another one of Grossman's favorite stories shows. One police officer, demonstrating admirable intentions, trained himself to disarm a close assailant with a lightning-fast snatch and twist. Over and over again, this cop would rehearse this disarming technique with friends and colleagues, snaring the gun, handing it back to his training partner, snaring the gun, handing it back. Soon, while on duty, the officer and his partner responded to a complaint about a threatening man at a convenience store. They entered the shop and split up to search for him. At the end of an aisle, the suspect suddenly swung around a corner and leveled a pistol at the cop, point-blank. Reacting instinctively, the cop seized the gun away with ninja-like quickness, and then — just as he'd rehearsed in practice — handed the pistol right back to the shocked suspect. The officer survived (his partner immediately swooped in and shot the criminal), but we can assume he changed his training routine drastically from then on.* John Leach tells a similar story in *Survival Psychology* of how we fall back on well-learned behaviors in a crisis. In Norway, one armed robber walked up to a young female bank teller, took out his gun, and commanded her to turn over her cash. Writes Leach, "The woman was momentarily so surprised that she could only respond the way she always had: she pointed to her left and said 'Cash? Next desk.' The gunman, it appears, was equally taken aback and fled empty-handed."

Experts like Grossman hope to instill solid instincts in their students' brains, but some inclinations just can't be untrained. Take the startle response, our built-in reaction to a sudden fright: we coil up, blink, curl our hands into fists, and so on. (This reaction, you'll recall, is what Darwin tried to resist in front of the puff adder cage.) Through the years, innumerable military and police instructors have tried to drill this response out of their charges, but to no avail;

---

*Grossman can regale a crowd for hours with similar stories. For instance, one department he knows of in an "unnamed West Coast city" would practice by pointing their index finger instead of a gun in training for shoot-outs — a practice they stopped after reports trickled in of officers pointing their fingers as guns in real emergencies.

it's simply automatic. So now trainers generally work around it. Because we all have a tendency to clench our fists when startled—a response that can have tragic consequences if an officer involuntarily squeezes the trigger of his pistol when someone surprises him—many cops get trained to keep their finger off the trigger until they really want to pull it. Likewise, because people plunged into a sudden crisis tend to unknowingly exacerbate their fear reactions by hyperventilating, combat trainers put their students in the habit of employing what they call "tactical breathing"—also known to laymen as "taking a deep breath." Whenever he can, Grossman teaches a simple exercise that he believes is as useful as anything known to man in counteracting fear: breathe slowly and deeply into your abdomen (not your chest) for four counts, hold for four counts, exhale for four counts, and repeat. Our breath, Grossman and others have observed, is like a bridge to the parasympathetic nervous system, which is in charge of calming the body down. By slowing the breath, we send a subtle message to the unconscious mind that everything is going to be okay.

An old military saw has it that "proper prior planning and preparation prevents piss-poor performance," and it's an axiom that stands true for each of us—not just grim-faced soldiers and cops. For all of the findings by those like Leach and Wigram showing the sheeplike reactions of people immersed in a crisis, an even greater body of research supports the idea that no matter what your personality type might be, training can save your skin when that tense moment comes. Remember, the Pan Am passengers and soldiers who shut down in response to the crisis didn't do so because that response was their destiny; they froze primarily because they hadn't prepared beforehand, leaving their overtaxed brains unsure of how to proceed. They didn't read the safety card. They didn't get the right kind of training. And so they met the life-threatening crisis befuddled and blank.

But as David Eccles of Florida State University's Human Performance Laboratory points out, individual dispositions tend to even out as people gain more training. In World War II, Eccles explains, the U.S. military tried to select ideal helicopter pilots according to psychological tests but soon found that this screening was a waste of time. "When they tested people off the street on tasks related to helicopter piloting, there seemed to be substantial differences in natural ability between people," Eccles told me. "But after all of them had done one thousand hours of training, they found that you could barely distinguish between them. No matter how they tried to select people, training was always more successful, because the training washes out the effects of whatever came before." Put another way, if you pick a hundred people at random and toss them into an emergency totally unprepared, they'll respond according to Leach's formula: ten or so will excel, ten or so will freak out, and the rest will just be stunned sheep. But take those same hundred people and offer them well-planned training and preparation—even silly things like asking them to *really* watch the safety video before their plane takes off—and their responses will change immeasurably. "Personality plays much less of a role in coping well with stressful situations if a person has received good training," Eccles explained. As Grossman might say, they may not necessarily rise to the occasion, but if they sink to a decent level of training, that's good enough.

Get Lost

For thirty years, until his retirement in 1995, the survival trainer Peter Kummerfeldt taught U.S. Air Force cadets how to stay alive in the wilderness if their planes went down in no-man's-land. He was gifted enough at this that he spent the last twelve of those years as the director of the Air Force Academy's three-week survival program, which all recruits had to complete. Yet for all of his expertise, Kummerfeldt's career as a survival instructor nearly ended with his first class.

Deep in the Sierra Nevada mountains one cool autumn day, Kummerfeldt was leading his debut class of eight Air Force soldiers along a ridgeline when several of his students began clamoring for a break. Kummerfeldt acquiesced, but on one condition: they could take a breather as long as they spent it showing him they could build a fire. "While this is going on," he recalled, "I'm looking down the side of this valley and there's this beautiful little stream down there—and I love to fly-fish." Kummerfeldt told his class he'd return in an hour and trudged down toward the distant creek. The fishing, he remembers, was great. His hour up, he repacked his rod and climbed back up toward his students—but they were *gone*. "That was not a big deal," Kummerfeldt told me. "But what became a big deal was my sudden realization that I couldn't find any evidence of there having been a fire. I thought the students had gone on with the hike, but then I realized that no fire means *I'm* the one who isn't where I thought I was: I'm lost." Kummerfeldt wasn't the only one lost, of course—he'd left his class lost in the woods as well. "I was absolutely panic-stricken," he said. "My first class, and I'd lost my students already! They're counting on me, the air force is counting on me, my career is over, I'm going to get court-martialed—this horrendous cycle of thinking began." Immediately, the terrified Kummerfeldt began sprinting blindly through the forest, screaming "Hey! Where are you?!"

But in the midst of this panic-fueled dash, a miracle happened: Kummerfeldt's training kicked in. "This little voice in my head said, 'Hey, dummy, go sit down,'" he recalled. "So I went over to a log and sat down." This is precisely what he advises lost hikers to do today—to get off their feet. Next, Kummerfeldt followed his second protocol: he took a drink of water. "A drink of water is incredibly calming," he explained. "When I was training airmen in the Vietnam era, every air crew member flew with a one-pint flask of water in his flight suit leg pocket. That was not there for hydration. That was to enable them to get past panic. I say it washes the taste of fear out of your mouth. Those two things, getting off your feet and having a drink of water, are two absolutely incredible lifesaving steps." Sitting on that log, re-collecting his wits, Kummerfeldt suddenly remembered something significant. *He had a map.* "The moment I gave myself a chance, my brain started functioning again," he said. Triangulating his position, he soon discovered just how far astray his fear-stoked sprint had taken him: he'd dashed *three miles* without even registering it. Kummerfeldt brushed himself off, straightened himself out, and hiked back to his students. They had no clue he'd been lost. His superiors didn't find out, either. "I didn't tell anyone in the air force this story until my retirement ceremony." He chuckled.

That day, Kummerfeldt learned something important. If he, an expert, could lose his cool so drastically when he found himself lost, then so could anyone. "Every damn article I've ever read about emergencies and survival says, 'Don't panic,'" he said. "Bullshit! You're going to panic. I don't care how good an outdoorsman you are, you've got to get past this belief that you won't panic. You're being set up there." (After all, no crisis-stricken person in history has ever said to himself, "Hmm...freaking out appears to be the most prudent course of action here.") It's what we *do* with that fear reaction that determines our fate, Kummerfeldt says. In his classes today, Kummerfeldt teaches a model called STOP to get through wilderness survival situations: *stop* what you're doing, sit down, and let the adrenaline subside; *think* about your options; *observe* your environment; and *plan* how best to survive. Then get to work. But above all, Kummerfeldt cautions, this strategy works only if people prepare before setting out. Have they brought the proper gear in case the weather turns sour? Have they packed any first aid? And have they learned the basics of wilderness survival? Not the how-to-skin-a-rabbit-with-your-car-keys stuff from TV, he adds, but essentials like how to keep warm and dry. Mostly, Kummerfeldt finds, we deal with the prospect of disaster by ignoring the very possibility that it might occur. "The likelihood of a survival experience happening to any given person is pretty remote, so it's easy to justify not preparing for it," he said. "But for those who do get lost, stranded, hurt, caught in weather or darkness, it's a very real situation."

On that frigid night in the Chosin Reservoir, preparation was the factor that saved Hector Cafferata and those he protected; he'd shot a rifle almost every day of his life, so when the moment of crisis arrived, firing with deadly accuracy was instinctive. Preparation saves Bernie Chowdhury every time he finds himself in an emergency a hundred feet below the sea. It saved the Pan Am passengers in Tenerife, even the tiny quantity they gleaned from reading a safety card. It saved Dave Grossman's cops, acting instinctively in sudden shoot-outs. And it has saved Kummerfeldt's many students who really did find themselves lost in the woods, with only the training and knowledge they'd acquired before departing to guide them safely through the crisis.

Kummerfeldt believes all of us should get lost at least once, if only to see what it's like to encounter that rare feeling of "your guts falling out" with terror. "Once you've been through it, you know what to expect, know you can get through it," he said. You learn about yourself. And you learn to be prepared—the most vital lesson of all. Said Kummerfeldt, "If you're not prepared for emergency, you're depending on luck. If you do prepare, then getting lost becomes just an inconvenient night out."

How to Be Afraid

In 1904, the great industrialist Andrew Carnegie heard the story of two men, Selwyn Taylor and Daniel Lyle, who died while trying to save workers caught in a massive coal mine explosion near his home city of Pittsburgh. Taylor and Lyle's sacrifice affected Carnegie so profoundly that he soon decided to found an organization to recognize the courageous deeds of such men — a nonprofit that thrives to this day. Since its establishment a century ago, the Carnegie Hero Fund Commission has awarded medals and financial gifts to more than nine thousand civilians who have put their lives in danger for the sake of others. Winning acknowledgment from the commission once takes incredible bravery. Many, like Taylor and Lyle, died in their attempts, snatching people from burning buildings or shielding defenseless victims from attackers. Winning the medal *twice*, then, requires a truly rare kind of valor. In fact, of all the medal's surviving recipients, only one man has been recognized by the Carnegie Commission twice for his heroism and lived to tell both tales. His name is Daniel Stockwell.

Stockwell was the first person I interviewed for this book, and through the dozens of subsequent conversations I had with athletes, soldiers, surgeons, and other remarkable people, he always stood out in my mind. This wasn't because of his bravado or bluster. Far from it: when I first spoke with him on the telephone, Stockwell's soft, gravelly voice sounded so quiet that I wondered if we had a bad connection. Stockwell didn't fit at all with the popular portrait of cool-headed heroism — the massively confident clutch quarterback or the steely-eyed fighter pilot. Instead, he was a reserved, recently retired principal at a small high school in rural New Hampshire. But more than anyone I've encountered since, Stockwell exemplified how to be afraid.

Stockwell performed his first act of heroism in 1963, when he was a junior at Bates College. One spring day, Stockwell and his girlfriend Merry — now his wife — attended the college's annual clambake on Fox Island, a rocky chunk of land off the coast of Maine. The Atlantic thrashed wildly that day, waves slamming into the island's stony cliffs. As Dan and Merry walked around, gazing at the violent ocean with astonishment, a girl ran up looking frantic. Two boys had fallen into the water, she shouted, and the current was sweeping them out to sea. They ran over to the site and several other students explained what had happened: a colossal wave had smashed into a fifteen-foot-high cliff, dragging one student into the turbulent sea; the other student dove in to save him, but fared no better himself. As Dan grimly began stripping down to his bathing suit, Merry beseeched him not to go into the water. Quickly, a few students cut down a volleyball net and tied it around his waist as a safety line. Dan had worked as a summer lifeguard before, but nothing could have prepared him for a sea this tempestuous. "I wasn't sure what I was going to do," Stockwell recalled. "The sight of the water itself was terrifying, but, I mean, the kids were drowning. I couldn't have lived with myself if I hadn't gone in."

Merely getting into the water over the sharp rocks and barnacles presented a challenge, and the freezing cold surf was so violent that Stockwell had to struggle just to see where he was. By chance, he spotted one of the students at the peak of a swell. After swimming over, he found the student unconscious and barely staying afloat. Stockwell put him in an armlock and held on tight as the other students reeled him in. On his arrival at the jagged shore, the barnacles tore his legs up badly; they were so swollen the following day that he almost couldn't pull on his pants. The student Stockwell hauled out made a full recovery. The other boy died. The college newspaper interviewed him about his heroism and the dean shook his hand, but no story about the tragedy ever came out. Everyone just wanted to forget it had ever happened.

That was Stockwell's first Carnegie Medal. Twenty-eight years passed. He and Merry started a family, raising two sons and a daughter. Stockwell became a teacher and then a principal, winning a job at Monadnock Regional High School in the small village of Swanzey, New Hampshire. And then, in the autumn of 1991, every parent's worst nightmare came to life in Stockwell's school.

One October morning, as a chattering crowd of students milled around the cafeteria waiting for class to begin, a sixteen-year-old dropout with long black hair walked in with a high-powered rifle hidden under his trench coat. When the bell rang, he fired his rifle in the air, and the shot ricocheted off a steel beam in the ceiling, wounding two students. Stockwell heard the shot from far off, and soon a flood of terrified kids surged in his direction. While Stockwell tried to sort out what was happening, the gunman took fifteen seventh-graders hostage in a classroom. A teacher told Stockwell that the teenager had gone down the hallway and proposed setting off the fire alarm to evacuate everyone. Here, Stockwell made his first cool-headed decision of the day. "I remember saying, 'No, don't touch that fire alarm,'" he told me, "because I was afraid the kids would go in his direction and make the problem worse." Instead, Stockwell posted teachers on each end of the corridor, to keep anyone from walking by, and told his secretary to get on the PA and instruct everyone to exit through the front door. He sent instructions to call the police, but he knew it could take a while for them to arrive. "It's a small town, and I've had to wait hours for a police officer," Stockwell explained. He knew *he* had to be the one to act.

Stockwell crept down the hallway and peered through a window into Room 73. The teenage gunman stood at the head of the classroom, reloading his rifle from a box of shells he had set on the teacher's desk. The students sat at desks or on the floor, some of them in tears. Stockwell walked back down the corridor and asked his assistant principal for information on the teenager, who had dropped out before Stockwell took the job. Then he marched back to the classroom, gathered himself, and knocked quietly at the door.

As soon as Stockwell opened it, the gunman aimed the rifle directly at him. He immediately noticed that the kid was trembling with fear; whatever plot he had entered the school with had gone wrong. Gently, addressing him by his first name, Stockwell asked if he could come in. The teen assented. Still speaking softly, Stockwell went into his pitch. "Whatever it is that you're planning to do, you don't need this many people," he told the teen. "Why don't you let them go? You've got the principal now, and you couldn't ask for a better hostage." The gunman turned to a girl who was crying. "Are you afraid?" he asked her. "Yes," she said, through tears. He considered for a moment and then told the kids they could leave. Stockwell's plan had worked: he had exchanged himself for the fifteen student hostages. "I felt pretty good for an instant—until I realized I wasn't out of it yet," he said.

For what seemed like an eternity, Stockwell stared down the

barrel of the agitated teen's rifle, the hammer still cocked, his trigger finger still shaking like a leaf. The terror Stockwell felt was like nothing he had ever experienced. "There were instances when I imagined hearing the click of the trigger and seeing the flash come out of the barrel, and I had to get that out of my mind," he said. "I just concentrated on the moment and didn't think about what could happen, because if I started thinking about the future I felt like I wasn't getting out of this." Through his fright, he focused on keeping the kid as calm as possible. He talked with the teen about what he hoped to achieve there, but Stockwell could tell he had no idea what he wanted; the teen just delivered bland diatribes about the evils of the government and the unfairness of life. Soon the gunman issued a halfhearted demand: he wanted a boom box. Despite his horror, Stockwell kept cool-headed here as well. "By then I'd noticed some police officers outside the building, and I'd also seen one walk by the door," he told me. "So I asked, 'Do you mind if I go to the window and tell the police what your demand is?' I could have gone to the PA near the door, but if I went to the window, he'd have to turn his back on the police officer in the hallway." The kid swiveled to face the window, but to Stockwell's chagrin, the officer didn't catch on. Stockwell returned to his seat, the rifle still leveled at him. They kept talking.

After a few more minutes, the teen told Stockwell he wanted to see a friend, and Stockwell duly walked to the window to relay his request. This time, the officers were ready. As Stockwell called out the demand, an officer burst through the door clutching his pistol in both hands. The teen swiveled around and suddenly moved to put the rifle barrel under his own chin. Another officer, sneaking in a side door with his shoes off, snatched the gun from his grasp and managed to subdue him. When Stockwell emerged from his terrifying ordeal forty-five minutes after first walking into the classroom, he had lost all sense of time. "I was concentrating so much on what was happening that I had no idea what time of day it was," he said. "I thought it was afternoon. I lost track of everything except the moment. My mind was on nothing but this kid with the rifle pointed at me." Years later, when the teenager got out of jail, Stockwell met with him and the kid told Stockwell he was amazed at how calm he seemed. Stockwell found this amusing. He had never been more frightened in his life.

That was Stockwell's second Carnegie Medal.

People often ask Stockwell what it feels like to be a hero, and he finds the question impossible to answer. Most assume that he must possess some supernatural psychological makeup that only heroes share. Yet in both his ocean rescue and his experience with the teenage gunman, Stockwell says he acted only out of necessity. "Those were my kids in there, and they were all alone," he told me. "Their parents sent those kids for me to take care of." What else was he supposed to do, if not act? Stockwell believes he might be cooler under pressure than some, but he says there's still nothing terribly unique about him. Whenever someone tries to heap praise on his actions — local newspapers, civic groups, even Hollywood producers with promises of hefty payouts for movie rights — he shies away. He did what he did because it was the right thing to do, no more, no less.

Sometimes, when English teachers at his old high school do class units on courage and heroism, Stockwell will go in and answer the students' questions. Without fail, they ask him if he was afraid. Of course he was, he responds. But he doesn't add the real truth to that answer, which is this: Dan Stockwell was, and always has been, more afraid than anyone imagined. Throughout his life, Stockwell has dealt with persistent anxiety and depression. "You know, I may appear solid as a rock, but I've had to overcome a lot of things," he told me. "I've been a pretty bad worrier even in the best of times." When he first started his teaching career, at a tough inner-city school in Connecticut, he was so frightened each workday that he routinely ate breakfast, immediately threw it back up, then went on to class. After he got promoted to principal, his anxieties deepened. As he explained it, "I was twenty-seven years old when I became a principal, and my first year I wanted to go home and stay under the coffee table. I was getting up six times in the night, just worrying about going to work."

Stockwell lived with fear for decades, fretting over what he called "shadows in the closet." He took medication, and read books on anxiety and depression. Even after decades as a principal, he still worried constantly—in fact, he fretted about precisely what ended up happening that October day. "One of the things I worried about was somebody coming into my building with a gun," he told me. "What would I do? There were different scenarios I would picture, and the only way I could think of for defending myself was with the archery equipment in the phys ed room." Over time, he has learned that the best way to deal with his anxieties is to accept them, coexist with them, and pay attention to the current moment. "I live day by day," he told me. "I don't live in the future, and I don't live in the past. You've got to live in the present."

Dan Stockwell is just as neurotic as any of us, just as troubled by the risks and uncertainties of life—and this is what makes him remarkable. His persistent fear, he now realizes, has always been beside the point. Said Stockwell, "You know, it has nothing to do with facing an issue. I may be that way, but *I'll face any issue*." This is the true lesson behind dealing well with fear. Stockwell, racked with terror in his courageous moments, stayed cool-headed and acted heroically not by fighting fear or banishing it but by working with it.

Once, Stockwell and his wife were driving to a hospital where Merry was scheduled to undergo major surgery. She was terrified. Her doctors believed she might have cancer. Her teeth chattered, and tears streamed down her face. "I'm sorry I'm not being brave about this," she apologized to her husband. Then Dan looked at her. You *are* being brave, he told his wife. Bravery isn't being *fearless*. Bravery, he said, is being scared and doing the right thing anyway.

That, I believe, is how to be afraid.

The Right Stuff, Unveiled

When we began this book, cool-headed and fear-hardy people seemed mysterious and elevated, like a breed apart from the rest of humanity. How, we wondered, did Vasily Arkhipov, entombed in a dying submarine with what may have been World War III erupting above him, keep his poise when everyone else wanted to launch a nuclear warhead? How does Kobe Bryant hit the clutch shot at the critical moment, with millions of eyes watching? How do firefighters and cops navigate through stress-inducing chaos day in and day out? Yet if grace under fire appeared to be beyond understanding when we first set out, I hope we've demystified the issue somewhat. And to show this, I'd like to tell you one last story.

In the early afternoon of May 15, 1963, the astronaut Gordon Cooper vaulted up through the Florida sky aboard the rocket *Faith* 7, commencing the most ambitious mission in the brief history of the American space program. Over the course of a day and a half, Cooper was to travel 600,000 miles on the cusp of space: twenty-two full orbits around earth, nearly triple the previous American record. Of the seven original Mercury program astronauts, Cooper considered himself uniquely qualified for such a challenging mission. A handsome Oklahoma boy and longtime test pilot, Cooper wasn't just the youngest and fittest of the Mercury astronauts—he was also the cockiest. In the 1983 film *The*

Right Stuff, based on the classic Tom Wolfe book, Cooper loves to ask, rhetorically, "Who's the best pilot you ever saw?" Cooper never distanced himself from the line; he really did believe he was the greatest thing ever to hit the sky. This attitude lent him an eerie cool in the face of danger. After Cooper strapped into his capsule that morning—a cabin so cramped with equipment, he later wrote in his autobiography, that "you didn't climb in so much as put it on"—he greeted the prospect of the longest ever manned American space mission quite unworriedly. Belted into the tip of a ten-story-tall Atlas rocket carrying 200,000 pounds of fuel and waiting to explode into liftoff, with the entire nation anxiously monitoring his every move, Cooper took a nap.

The flight went perfectly for the first eighteen orbits. Cooper ran experiments in the weightless environment of his cabin, took photos out of his tiny porthole, ate cubes of peanut butter, and even became the first American to sleep in space. But on his nineteenth orbit, trouble appeared. It started with a green light that lit up unexpectedly on his instrument panel. This light was meant to indicate that the pod was decelerating and beginning to fall out of orbit; it wasn't supposed to turn on until reentry. Cooper was briefly out of radio contact at the time, and puzzled NASA officials (whose comparable indicator light had also illuminated) believed Cooper had initiated his return flight early. He hadn't. While both parties independently tried to figure out what was going on, *Faith* 7's entire electrical system began shorting out. Cooper lost telemetry. He lost his cooling and oxygen purification systems, meaning that carbon dioxide would soon start building up in the cabin. Then the gyroscopes failed. And the onboard clock. As Cooper infamously understated the situation over the radio (which continued working because it was on its own battery), "Things are beginning to stack up a little." On his twentieth orbit, NASA informed him that *Faith* 7 was experiencing a "total power failure." Wrote Cooper years later, "We all knew, without saying, that we were experiencing the worst systems failure in the more than two years that America had been sending men into space. *I was on a dying ship.*"

Because his automatic control systems had been rendered useless, Cooper would have to perform a feat that had never before been accomplished: he would need to steer his way back through earth's atmosphere manually—and he'd have to do it without help from his instrument panel. I'll let Cooper himself explain just how daunting a challenge this presented:

> With my gyros out, I would have to establish my spacecraft's angle of attack using only the horizon as my reference point. Then, manipulating the hand-grip controller located next to my seat, I'd have to fire the retro rockets at precisely the right moment and hold the spacecraft steady on all three axes— pitch (vertical), roll (lateral), and yaw (side to side). When the electronic damping system was working, these corrections were automatic, offsetting for the thrusters so that the spacecraft didn't start oscillating wildly. Now, I'd have to damp out any oscillation as soon as it happened or the spacecraft would quickly spin out of control and rapidly disintegrate in the atmosphere.

In other words, if Cooper didn't fly the ship just right, it would enter a death spiral and burn into cinders within seconds. Raising the difficulty level even higher, the temperature in the *Faith* 7 cabin had already reached 130 degrees Fahrenheit. Cooper felt himself panting, a first sign of carbon dioxide poisoning. It was time to fly.

At 6:03 p.m., soaring over Shanghai, Cooper fired his retro rockets, slowing *Faith* 7 from 17,500 miles per hour to a leisurely 1,200 miles per hour and dropping the ship from orbit. Immediately, he felt the craft destabilizing. He had to compensate constantly, always keeping his heat shield positioned at the right angle as the earth's atmosphere slowed his fall. Encased in a flurry of superheated ions, he entered a communications blackout. For fifteen minutes, with heat and carbon dioxide building in the cabin, Cooper kept flying straight. As he pulled the craft's parachutes manually, he paid special attention to his exact pitch and timing; Cooper had made a bet with Wally Schirra, the astronaut from the previous Mercury mission, over who could splash down closest to their recovery vessel, an aircraft carrier called USS *Kearsarge*. Even now, Cooper aimed to win it. Schirra had landed six miles from the carrier and was quite proud of his precision. But Gordon Cooper, piloting a wounded spacecraft under extreme stress, receiving no help from his instruments, did him one better. He landed five miles away from the *Kearsarge*. "If it hadn't been for the wind causing drift, I believe I would have landed right square on the carrier deck," Cooper later bragged.

Cooper's performance that day surely ranks in the pantheon of cool-headed accomplishments under pressure. With the entire country watching him and his very life on the line, he overcame physical stress and his own intense fear—and by all physiological reports, he was plenty afraid—to accomplish a feat that had never been seen in spaceflight. For most of us, such an achievement is beyond imagining, and so we see Cooper as someone who stands above mere mortals. Maybe he was just born that way, we might think. But while Cooper's innate psychological disposition surely played a role in his accomplishment, it's not even half of the story. The majority of what went into Cooper's cool-headedness was the product of factors he developed over a lifetime, factors we can identify and quantify. After all we've learned about fear, pressure, and stress, let's peel back the layers of the onion a bit. What *really* went into Cooper's grace under fire aboard *Faith* 7?

- *Confidence:* Cooper, as we've seen, had a massive ego, and such confidence—built up through many small goals successfully accomplished over time—buoys us in fearful situations; it makes us feel we're equal to a challenge instead of under threat. "If I sometimes acted as if I didn't think anyone could outdo me in the air, well, that's how I felt," Cooper once wrote. "Modesty is not the best trait for a fighter pilot. The meek do not inherit the sky."

- *Training:* At the time, the Mercury astronauts were some of the most hyperprepared people who had ever walked the earth. They spent countless hours in a training capsule, rehearsing every conceivable scenario so they'd be able to respond to it automatically. Wrote Wolfe in *The Right Stuff,* "The idea was to decondition the beast completely, so that there would not be a single novel sensation on the day of the flight itself." Everything Cooper did that day he'd practiced hundreds of times before—even piloting the capsule through reentry with no power.

- *Locus of control:* We all meet a frightening situation with less anxiety if we believe we can influence the outcome, and the Mercury astronauts, as former test pilots, belonged to a culture that took this to its extreme. Wolfe writes that whenever one of their colleagues died in a crash, they would all get together and agree that it was tragic—but how could the pilot have failed to do x, y, or z? They pinned each death on controllable factors *they* would never overlook, building a sense of power over their fate.

- *Tolerance for uncertainty:* Still, the astronauts knew from their test-piloting days that equipment failures are the nature of the business, and they had to face up to that uncertainty in order to do their jobs. In fact, before John Glenn made the first manned American spaceflight, the type of rocket that propelled him had unexpectedly blown up a total of thirteen times in trials. Being an astronaut was risky, but they accepted that risk.

- *Humor:* Even with his ship falling apart around him, Cooper could lighten his predicament with a joke. At one point, he told mission control that he'd lost all electrical, cooling, and oxygen filtration systems, but added, "Other than that, things are fine."

- *Task focus:* Cooper never allowed himself to become preoccupied with his fear or to lapse into worries about what might go wrong; instead, he focused on what he needed to do *in the present moment*. At a press conference after he landed, someone asked him if he ever feared for his life, and the phrasing of Cooper's answer is telling. "I felt my options were rapidly decreasing," he said. At all times, he zeroed in only on what he could do to help himself.

By now, the overarching point should be clear. How did Cooper really stay cool under life-threatening pressure? The same way everyone else we've met did it: by working *with* fear instead of needlessly wasting energy fighting it, by focusing on what needs to be done instead of on worries, and by taking action. Cooper wasn't necessarily good at *not feeling afraid;* the onboard physiological monitoring equipment actually showed that he was terrified. What Cooper had become good at was making this fear *irrelevant*. When the director Ron Howard was interviewing astronauts before making his film *Apollo 13*—which chronicled another space mission gone awry—he noticed much the same thing. "It's not that they ignore fear," Howard has said, "but they have dealt with it so many times, they actually come to expect moments where the natural inclination is going to be to panic and they have trained themselves...to feel fear, even embrace it. And then, almost like tai chi or something, they sort of let that go through them and work right through the problem." The astronaut Buzz Aldrin would concur.

Opining on fear in a 2003 *New York Times* editorial, Aldrin wrote, "I don't think anybody—astronauts or otherwise—is born with some kind of right stuff. It's something you work into." (Less thrilling, for me, was Aldrin's addendum to this thought: "Or else you wash out of flight school and get a job as a journalist.")

From astronauts to athletes, soldiers to surgeons, the sought-after secret to dealing well with fear remains the same: it's not whether you feel afraid that matters, but how you react to that fear. Look at Laurence Olivier, who was beset with terror even as he delivered some of the finest stage performances of the twentieth century. Consider Dennis Eckersley, who became one of the greatest clutch pitchers of his generation while toting his anxiety onto the mound with him every night. Think of Al Haynes, who figured out how to pilot his disabled jet onto an Iowa runway under intense stress without the ability to steer. Imagine Hector Cafferata, protecting his fellow soldiers through the night, so amped up on terror that he didn't notice he wasn't wearing boots in the snow. And, of course, remember Dan Stockwell, lifelong worrier, as fearful as any of us, twice pushing his way through extreme fright for the sake of others. All were afraid. All performed at the highest level imaginable.

What unifies these people isn't their fearlessness but their attitude toward their fear. All of them followed a road that's available to any of us, at any time. They accepted how they felt, focused themselves on doing what was right, and got on with it.

So let's get on with it.

Applying Nerve

Thus far in *Nerve*, I've focused almost exclusively on showing how fear and anxiety aren't villains we must vanquish but benign forces we can accept and live with. If I've had a secret agenda here, it's been to get us comfortable with this welcoming, harmonic view of fear before presenting a slew of tactics for managing it. But that said, we can all use practical tips in our quest to relate better with our fears, stresses, and worry-laden minds—and after several years of neurotically obsessive research, I've built up a short list of the most effective, proven methods that psychology and neuro-science have to offer. I can't promise any panaceas or instant fixes to eradicate anxiety forever, demolish stress, or ensure clutch per-formance, and neither can anyone else. Fear is a fact of life. All we can do is learn how to be afraid in the right way. So herewith, a dozen quick tips to get us started.*

Breathe. This pointer might sound preposterously obvious, but if you're feeling stressed or anxious, chances are you're breathing quickly and shallowly into your chest, or even *holding* your breath. This just perpetuates a fear or stress reaction; it sends a message to the body that something's amiss. When we're calm (or asleep), we breathe far differently: slowly and deeply, drawing air into the gut. By consciously controlling our breath, we can inform our parasympathetic nervous system that things are okay, lowering our heart rate and taking fear down a notch. Here's the "tactical breathing" method that the combat psychologist Dave Grossman teaches to his students, for use on gritty battlefields and in tense university testing halls: slowly draw air through your nose down into your abdomen for four leisurely counts (you can place a hand on your stomach to make sure you're breathing in correctly), hold for four counts, exhale through your mouth for four counts, hold again for four counts, then repeat as necessary.

---

*For more advice from the pros, please see the Suggested Reading sec-tion in the back of the book.

Put your feelings into words. As we learned in Chapter Two through Matthew Lieberman's labeling research, talking or writing about an emotion like fear helps the brain to process it behind the scenes; it allows the mind to disambiguate (i.e., sort out) thoughts and feelings instead of just churning over them repeatedly. "People write in journals and talk to friends not realizing that it's changing their emotions, but it is," Lieberman told me. Too often, we bottle up our fears and anxieties, but speaking compassionately and honestly about emotions, without judgment or self-blame, helps us come to terms with them. This is why talk therapy is so effective, but writing expressively about fears in a personal journal works too. As the novelist William Faulkner once wrote, "I never know what I think about something until I read what I've written on it."

Train, practice, and prepare. We've seen this theme come up time after time, on game show sets, in trauma wards, and aboard space capsules. Whether you want to make better decisions under stress, handle life-threatening situations with composure, or perform your best when pressure hits, training is the only reliable way to ensure success; through repetition and experience, you program yourself to do the right thing automatically. If an upcoming presentation has you anxious, rehearse it under realistic, flexible conditions until delivering it becomes routine. If you're worried about how you'd react if your building caught fire, learn the right exit routes and practice taking them. Keep the U.S. military's eight *P*s in mind: "Proper prior planning and preparation prevents piss-poor performance."

Redirect your focus. From nervous student test-takers to anxious athletes to jittery actors, the culprit in cases of meltdown under pressure isn't *fear* but misdirected focus: we turn our attention inward and grow preoccupied with worries about results, which undercuts our true abilities. Clutch athletes and cool-headed heroes concentrate on the *present moment* and on *the task at hand*, a habit we can all develop through practice. Meditation certainly builds concentration, but we don't necessarily need to go that far; psychologists say that even pausing a few times a day and being present for a moment with what's going on around you (rather than with the monologue in your head) can help you to better inhabit the current moment. This requires patience, though: the chattering of the monkey mind loves to intrude, which brings us to...

Mindfully disentangle from worries and anxious thoughts. We know now that worry does us no good, yet trying to *stop* our fretting altogether is well nigh impossible. But luckily, we don't have to manhandle our anxious thoughts or win debates with our worries; we just need to develop some distance from that fretful (not to mention harmless and frequently misleading) voice in our heads. Evelyn Behar, the worry expert, suggests two paths for detaching from this internal chirping. One is to take the mindfulness route: the more you learn to simply watch your worries and let them coast by without getting entangled with them, the more you see them and their predictable patterns as if from far above. "Or," Behar continued, "you can postpone worry. You write a worry down and agree that later on you can worry about it for thirty minutes, which frees you up to focus on the moment." Over time, she says, you learn to better manage your anxious thoughts and they no longer run the show.

Expose yourself to your fears. We've gone over this extensively, but it bears repeating. If you want to remain locked into a fear indefinitely, then by all means, avoid the situations that make you anxious. But if you want to give your amygdala a chance to get over a fear, you must expose yourself to the things and ideas that scare you. Whether you're dealing with a slight aversion or a full-bore anxiety disorder, *avoidance*—not fear—is the scoundrel holding you back. Behar offers a good rule of thumb here: if anxiety is stopping you from doing something that isn't objectively dangerous, do it anyway. Exposure is no guarantee of future serenity, mind you; if you meet a fear too confrontationally, cursing fate and vilifying your anxiety, you won't give your brain much space to learn that you can handle it. A good therapist can help manage the exposure experience wisely, and if she uses the learning enhancement drug d-cycloserine, she may even be able to speed the process up. Just get in the habit of moving *toward* your fears rather than running away. When you do so, even "failures" become successes, each exposure two steps forward to one step back.

Learn to accept uncertainty and lack of control. Anxiety and stress feed on our negative response to feeling uncertain or powerless over the future—and when people (such as hypochondriacs) quest after unattainable certainty and control, they end up making things far worse. So instead of tilting at windmills, anxiety experts like Robert Leahy suggest taking a hint from the well-worn Serenity Prayer, which aspires to "the serenity to accept the things I cannot change, the courage to change the things I can, and the wisdom to know the difference." When we're troubled about something uncertain or uncontrollable—possible illness, relationship doubts, financial worries—Leahy recommends a simple practice to help us accept reality. Suppose you're worried you might be laid off from your job. Leahy says that if you bask in your uncertainty (that is, expose yourself to your fear about the future), repeating the distressing thought *It's possible I could be laid off* to yourself without resisting your anxious emotional reaction, then you (and your amygdala) will eventually begin habituating to it. With enough exposure, the idea loses its power and becomes almost dull.

Reframe the situation. When the procession of negative biases and anxious thoughts starts marching through our heads, we always have an important choice to make: do we buy into a falsely pessimistic interpretation of what's going on, or do we learn to see things differently? "I like to say you can make an emotional molehill into an emotional mountain, which is what people do all the time," said the Columbia psychologist Kevin Ochsner, who studies how we reappraise situations to adapt our emotional responses. When anxious biases float in, Ochsner stresses the importance of recontextualizing: staying grounded in reason and reminding ourselves of the (doubtlessly more positive) reality of our situation. Social phobics, for example, tend to assume that everyone at a party is watching them; nervous performers think everyone can see their butterflies; neurotic people like myself see danger in change and uncertainty. Reframing things with a more optimistic and realistic spin — no one's watching the social phobic, no one can see the performer's nerves, and change could easily be *positive* — lets us keep our fears in the right perspective. "We do this all the time in everyday life, right?" said Elizabeth Phelps, the NYU reappraisal expert. "When you change the way you appraise a situation, you change your emotional response to it."

Joke around. As the humor researcher Rod Martin explained in Chapter Three, thinking playfully or joking in a stressful situation helps us break out of a negative point of view and see things differently. (This, you'll notice, could be considered a subtle and useful variation on the reframing method above.) By poking fun at life's occasional grimness, we neutralize its venom and lift ourselves above it. If you're not a natural comedian or find your-self in a less than jocular mood, that's not a problem. Humor is helpful because it lets us play around with concepts and points of view, so anything you can do to find the ridiculous, less serious side of a rough situation will lighten the mood. (Stay within the bounds of good taste, of course; no one's suggesting you visit the trauma ward and launch into a standup routine on hospital food.)

Build faith in yourself. This tip might sound a bit like inspira-tional drivel, but developing confidence that you can handle intense fear and stressful predicaments is absolutely vital. Confi-dence transforms dangerous-seeming threats into challenges we can overcome, it gives us a sense of control over our fate, and it keeps us plugging away at problems until we find a solution instead of just giving up. We don't gain this confidence in one miraculous snap of the fingers, obviously; we do it step-by-step, through a series of small successes that gradually expand our comfort zone and our self-belief. At the risk of sounding sappy, you might doubt your ability to cope with fear or a difficult situation, but believe me, you can—you just don't know it yet. Remember, worry research shows that people handle worst-case scenarios far better than they ever expected, and therapists like David Barlow like to plunge their clients into deep terror to show them reserves of strength they didn't know they had. And in addition to building confidence through fear exposure, we can also do it through the ways that we talk to ourselves and handle worrisome visions of the future. Here's a useful practice: next time you imagine something you fear coming to pass, visualize yourself not enduring it misera-bly or falling apart but *coping with it well*, demonstrating grit and resilience. Because in all likelihood, that vision of coping is what would happen.

Keep your eyes on a guiding principle. Fear, anxiety, and stress can make the universe seem chaotic and bewildering, so it's always helpful to have a compass to steer you through the maelstrom. Dedication to a higher purpose—be it spiritual belief, altruism, or personal goals and principles—helps keep us afloat and pointed in the right direction when everything appears scary or hopeless. The psychologist Steven Hayes, who created acceptance and commitment therapy, says devotion to personal values is a crucial part of learning to live with anxiety and stress. Hayes believes our emotional pain helps highlight what's really important to us. "If you flip anxiety over, it tells you what you care about, what your values are," he told me. A man who fears public speaking, for instance, might really be anxious about fumbling in public because he cares so deeply about connecting with others, and thus hidden within his fear is a positive value on which he can focus. Exposing ourselves to fear for the sake of what we believe can utterly transform our experience of it. The philosopher Friedrich Nietzsche put it best when he wrote, "He who has a *why* to live for can bear with almost any *how*."

Open up to fear unconditionally. It sounds simple enough in principle, but perfecting this important practice can be the work of a lifetime. If I were to delve into all of the stumbles, blunders, and false starts I've gone through in my own journey to deal better with fear and anxiety, I could fill another book or ten: fighting anxiety, trying to control it, avoiding it, seeking to figure it out intellectually, blaming myself for feeling it, and so on. After making enough mistakes, you eventually realize that you've been treating your fears like an absolute jerk. There's nothing wrong with feeling anxious, ever, over anything at all. Fear and anxiety are part of who we are. Once we drop the pointless, wrongheaded routine about needing to get rid of them, we can carry fear and anxiety around with us through life like friendly companions. Instead of battling fear, we just let it happen, and when the fight against it dissolves, so does the torment. We slowly learn to live in harmony with fear, anxiety, and stress, expecting them to show up and welcoming them when they do. And then a problem that once seemed so horrific and intractable becomes so simple—and really, not a problem at all.

The Path

As I've learned firsthand, there is no steadfast rule or polished gem of knowledge in existence, here or elsewhere, that will instantly make you calm or cool-headed in the face of fear. There is no cure for fear, nor should there be one. Fear is *good*. It protects us, gives body and character to existence, points us in the direction of what we value. We couldn't make decisions or perform our best or even survive without it. Fear can be uncomfortable, sure, but this doesn't prevent it from being an unconditionally positive force in our lives. Rather than agonize over our anxieties, we would be wiser to greet fear like a trusty sidekick. This new friendship with fear and anxiety doesn't come from a psychological switch we flip or an instant fix we apply. It's not a set destination we can reach and then stop making an effort. It's a path we walk every day, always progressing, never arriving.

I'll use a parallel to illustrate what I'm talking about here. For decades, the Stanford psychologist Carol Dweck has researched how our theories of intelligence shape our performance on mental tasks—or, as she puts it, she studies "beliefs that make smart people dumb." Broadly speaking, Dweck says we have two conceptions of intellectual ability. Some of us believe intelligence is a "fixed trait" like eye color or height; either you're smart or you're not, full stop. Others see intelligence as a malleable "repertoire of abilities that can be cultivated over time"; these people think that how "smart" you are is mostly a product of the effort you put into developing your own mind. What Dweck has found over and over again is that regardless of a person's innate intellectual gifts, the belief in intelligence as a fixed trait is actually *toxic* to their performance—it makes smart people dumb. Those who think of brainpower as rigid and preset believe so strongly in the power of things like individual test results to tell them how smart they are, Dweck saw, that any setbacks made them lose heart and quit. If they got a D on the math test, then a D-student was what they were; why even try to improve? Their belief that they couldn't change made them stagnate. On the other hand, the students who thought of their intelligence as something they could guide and nurture excelled in Dweck's studies. When these students did poorly on tests, they responded with greater effort. They tried a broader array of subjects and thrived on challenges, even if they faltered at first. So what if they tripped up in the short term? In the long term, their sustained effort made them smarter and smarter.

The same lesson applies to dealing with fear. In one conception, a privileged few among us are naturally great at handling fearful situations, while others are fated to flounder forever. But this just isn't true. In reality, our ability to deal well with fear is largely a product of changeable skills that we can develop over time. Yes, many of us have unfavorable genetic predispositions to contend with, and yes, our amygdalae have minds of their own. But the vast majority of what goes into handling fear, pressure, and stress with poise is directly under our control—it's the result of the path we walk through life. Do we avoid the things that scare us, or do we make a practice of seeking out and experiencing our fears? Do we allow ourselves to worry endlessly about imponderables, or do we make the deliberate effort to become more mindful and engaged with right now? Do we crack a joke when times are tough, build faith in our ability to navigate intimidating situations, teach ourselves to live with uncertainty, cultivate flexibility in an ever-changing world? And most important of all, do we learn to accept our fear instead of pushing it away? So the measure of our ability to deal well with fear isn't *whether* we get afraid, but how we *connect* with that fear. In the end, this is the kind of "nerve" that matters most: the nerve to open up to fear, work with it, and do the right thing regardless of how we feel.

This path isn't always easy. Welcoming our fear, doing what scares us—these are often the *last* things we want to do. The trail can be rocky. The hike takes work; no one gets to just sit on the path and make progress by magic. Life grows no less uncertain, no more controllable. We might become calm and placid, and we might not—that's not really the point. Instead of wishing that the road will someday be smooth and easy, we hope that the experience will strengthen us for future peaks and valleys. I can be as neurotic and worry-laden as they come, and I expect to stumble doggedly along this path for the rest of my life. I've spent several years researching fear, anxiety, and stress now, and while I feel empowered by what I've learned, I realize that the knowledge itself takes me only so far. There's a reason that people who have never swum a single stroke don't read a stack of books on swimming and then hop into the English Channel expecting a trouble-free crossing to the other side. Knowing what to do on an intellectual level is the easy part. The hard part is actually doing it. It's standing up in front of that terrifying crowd, getting onto that airplane, stepping into that classroom with the rifle-wielding teenager, and being there in harmony with your fear. Walking that path is difficult, but we must walk it anyway. Because the path of fear is the path of life.

Acknowledgments

In Shakespeare's *The Merchant of Venice*, the embittered money-lender Shylock famously tries to extract a pound of flesh from the merchant Antonio to repay a debt. After writing this book, I think I could tell Antonio a thing or two about what the procedure would have felt like. I carried a hefty load of worry and anxiety into this project, and at times it seemed as if I were carving this book from my own gut, slowly and painfully. I wish I could say I was exaggerating for humor's sake, but writing this book has been the hardest thing I've ever done. In the process, I sometimes treated Chapter Two's eight anxious mistakes like a psychological Whitman's Sampler, gorging on the tempting morsels and then feeling guilty and unfulfilled afterward. I believe the clinical term for this is "learning the hard way." So, for better or worse, I do feel like there's a pound of my flesh in this book.

In Shakespeare's play, Antonio is rescued from certain doom through the assistance of kindly allies—particularly a beautiful, gracious damsel—and to this I can relate as well. I would have had neither the courage to begin this book nor the wherewithal to see it through without the compassionate support of my lovely girl, Libby Tucker. There were times during the manuscript-writing process when I was hardly a bundle of joy to live with, but she always stuck with me, and she helped me to be a better man. For that, and for so much more, she has my eternal love.

I also owe a huge debt to my mom and dad, who provided every shred of support they could manage. I've always known that either of them would step in front of a moving bus to help me, and that means far more to me than I could ever express. I am likewise grateful to Gina, Lauren, Laurel, and Tim, who all offered invaluable moral assistance when I needed it. Zach Dundas was my comrade-in-arms at the office, and his encouragement, humor, and distinctive Victorian cowboy style helped me keep my head screwed on more or less straight. Wilson Vediner and Carissa Wodehouse listened to me fret in circles on countless occasions and deserve special merit badges for their unflagging aid. My thanks as well to Nigel Jaquiss, Tom Bissell, Francesca Monga, Liz Nagle, Vijay Shankar, Janette Fletcher, and Anthony Georgis.

At Little, Brown, I had the incredible good fortune of working with not one but two gifted editors, Tracy Behar and Geoff Shandler. Geoff deserves extra kudos for two reasons: first because he helped me develop the idea for this book, and second because he tried to take a call from me when he was *lying in a hospital bed*, until his wife snatched the phone from his grasp. Also, many thanks to Liese Mayer and Alison Miller. And at Trident Media Group, my deepest appreciation goes as always to unflappable superagent Melissa Flashman, who soothed my esoteric worries, solved my dilemmas, and boosted my spirits with her frighteningly boundless enthusiasm. I have no idea how other people write books without Mel as their agent; I certainly don't recommend it.

Finally, my undying gratitude, love, and friendship go out to Ember, the Professor, and the rest of the team. I hope I've done you proud.

SUGGESTED READING

F ear, anxiety, and stress are each colossal subjects, and while I've tried to pack plenty of useful information into *Nerve*, I could never dream of being totally comprehensive. For those readers who would like to explore the topics we've covered more deeply, here's a quick list of useful and trustworthy resources.

If you want effective, research-backed guidance on how to deal with general anxiety and phobias, start with...

Mastery of Your Anxiety and Panic: Workbook (4th edition), by David H. Barlow and Michelle G. Craske (Oxford University Press, 2006).

The Mindfulness and Acceptance Workbook for Anxiety: A Guide to Breaking Free from Anxiety, Phobias, and Worry Using Acceptance and Commitment Therapy, by John P. Forsyth and Georg H. Eifert (New Harbinger Publications, 2008).

If you'd like a more thorough look at how to address worry and rumination in particular, try...

The Worry Cure: Seven Steps to Stop Worry from Stopping You, by Robert L. Leahy (Three Rivers Press, 2006).

The Worry Trap: How to Free Yourself from Worry & Anxiety Using Acceptance & Commitment Therapy, by Chad Lejeune (New Harbinger Publications, 2007).

If you want to know every little thing about the amygdala and the neuroscience of fear, check out...

The Emotional Brain: The Mysterious Underpinnings of Emotional Life, by Joseph LeDoux (Simon & Schuster, 1996).

If you'd like an enjoyable, detailed look at the full story on stress, pick up...

Why Zebras Don't Get Ulcers (3rd edition), by Robert M. Sapolsky (Holt Paperbacks, 2004).

If a more spiritual, sagacious exploration of fear and life's challenges appeals to you, try...

When Things Fall Apart: Heart Advice for Difficult Times, by Pema Chodron (Shambhala Classics, 2000)—or, really, anything by the incomparable, invaluable Chodron.

If you're searching for a clearheaded guide to dealing with performance anxiety and stage fright, check out...

Notes from the Green Room: Coping with Stress and Anxiety in Musical Performance, by Paul G. Salmon and Robert G. Meyer (Jossey-Bass Publishers, 1998).

In the SpotLight: Overcome Your Fear of Public Speaking and Performing, by Janet E. Esposito (In the SpotLight, 2005).

If you want the whole, unvarnished truth on pressure performance in sports (and don't mind reading a textbook), look up...

Foundations of Sport and Exercise Psychology (4th edition), by Robert S. Weinberg and Daniel Gould (Human Kinetics, 2006).

If you're planning on fighting crime or visiting a war zone and need a fuller picture of how the mind responds to lethal encounters, try...

On Combat: The Psychology and Physiology of Deadly Conflict in War and in Peace, by Dave Grossman and Loren W. Christensen (Warrior Science Publications, 2008).

If you'd like to know more about surviving emergencies, disasters, and all varieties of catastrophe, pick up...

Deep Survival: Who Lives, Who Dies, and Why, by Laurence Gonzales (W. W. Norton, 2004).

The Unthinkable: Who Survives When Disaster Strikes—and Why, by Amanda Ripley (Crown, 2008).

NOTES

INTRODUCTION: *The Closest Call*

For the details about the story of the Soviet submarine *B-59*, I am primarily indebted to Svetlana Savranskaya's excellent historical detective work in her piece "New Sources on the Role of Soviet Submarines in the Cuban Missile Crisis," *Journal of Strategic Studies* 28, no. 2 (April 2005). Also see Ryurik A. Ketov, "The Cuban Missile Crisis as Seen Through a Periscope," *Journal of Strategic Studies* 28, no. 2 (April 2005); Michael Dobbs, *One Minute to Midnight: Kennedy, Khrushchev, and Castro on the Brink of Nuclear War* (New York: Knopf, 2008); Alexander Mozgovoi, *Cuban Samba of the Foxtrot Quartet: Soviet Submarines in the Caribbean Crisis of 1962* (Moscow: Military Parade, 2002); the History Channel program *History Undercover: October Fury*; and Marion Lloyd, "Soviets Close to Using A-Bomb in 1962 Crisis, Forum Is Told," *Boston Globe*, October 13, 2002.

The statistics comparing American and Mexican anxiety disorder rates come from the multinational 2002 World Mental Health Survey, information about which can be found at www.hcp.med.harvard.edu/wmh/index.php. The $300 billion cost figure for stress ailments is from John Schwartz, "Always on the Job, Employees Pay with Health," *New York Times*, September 5, 2004. Data on drug prescription trends between 1997 and 2004 can be found at www.meps.ahrq.gov/mepsweb/data_files/publications/st163/stat163.pdf. The Robert Leahy quotation comes from his *Psychology Today* blog Anxiety Files, which you can read at www.psychologytoday.com/blog/anxiety-files.

For a preposterously detailed look at Neil Armstrong's lunar landing, see David A. Mindell, *Digital Apollo: Human and Machine in Spaceflight* (Cambridge, Mass.: MIT Press, 2008).

Chapter One: *Your Second Brain: Exploring the New Science of Fear*

For dozens more firsthand survival stories from the 2004 Indian Ocean tsunami, see phukettsunami.blogspot.com. Also, the Raderstorfs' own contemporaneous recollections of the tsunami can be found at raderstorfwwa .blogspot.com/2004/12/alive-and-well-in-bangkok.html and raderstorfwwa .blogspot.com/2005/01/and-now-for-rest-of-story.html.

For Darwin's full account of his bout with the puff adder, see his book *The Expression of the Emotions in Man and Animals* (New York: Penguin Classics, 2009).

The science and evolutionary history of fear have been covered in several books, but for my money, the most useful and readable of these is Rush W. Dozier's *Fear Itself: The Origin and Nature of the Powerful Emotion That Shapes Our Lives and Our World* (New York: St. Martin's Press, 1998).

To see a highly entertaining video of the hognose snake playing dead, head to www.youtube.com/watch?v=7nScxF8vGw0.

For more on Livingstone's tonic immobility in the lion's jaws, see Dozier, *Fear Itself*.

Joseph LeDoux's innovative research on the amygdala and the neuroscience of fear is covered in loving, exacting detail in his two books *The Emotional Brain: The Mysterious Underpinnings of Emotional Life* (New York: Simon & Schuster, 1996) and *Synaptic Self: How Our Brains Become Who We Are* (New York: Penguin, 2002), the latter of which is rocky territory for laymen to navigate. While the vast majority of the LeDoux quotations come from my interview with him, I've also drawn a couple of them from *The Emotional Brain*.

I first found the story of Édouard Claparède and his memory-impaired patient in Steven Johnson, "Emotions and the Brain: Fear," *Discover*, March 2003.

On the University of Virginia snake grid study, see Vanessa LoBue and Judy S. DeLoache, "Detecting the Snake in the Grass: Attention to Fear-Relevant Stimuli by Adults and Young Children," *Psychological Science* 19, no. 3 (March 2008).

On the primate fear modeling studies, see Susan Mineka et al., "Selective Associations in the Observational Conditioning of Fear in Rhesus Monkeys," *Journal of Experimental Psychology: Animal Behavior Processes* 16, no. 4 (October 1990).

For more on the Little Albert experiment, see Joanna Bourke, *Fear: A Cultural History* (Washington, D.C.: Shoemaker & Hoard, 2006).

A useful discussion of "flashbulb" memories and the *Challenger* experiment can be found in Elizabeth Phelps, "Emotion and Cognition: Insights from Studies of the Human Amygdala," *Annual Review of Psychology* 57 (2006).

The Pitman propranolol study is in Roger K. Pitman et al., "Pilot Study of Secondary Prevention of Posttraumatic Stress Disorder with Propranolol," *Biological Psychiatry* 51, no. 2 (January 15, 2002). That study, plus several other cutting-edge ideas for manipulating fear memories, is discussed in Marc Siegel, "Can We Cure Fear?," *Scientific American Mind*, December 2005.

CHAPTER TWO: *The Worry Trap: Eight Awful Ways (Plus a Few Good Ways) to Deal with Fear and Anxiety*

For a more thorough look at the history and roots of hypochondria, see Susan Baur, *Hypochondria: Woeful Imaginings* (Berkeley: University of California Press, 1988). Additionally, Jennifer Traig's hypochondria memoir *Well Enough Alone: A Cultural History of My Hypochondria* (New York: Riverhead, 2008) and Gene Weingarten's humor book *The Hypochondriac's Guide to Life. And Death* (New York: Simon & Schuster, 1998) provide entertaining and profound firsthand viewpoints on health anxiety.

The $20 billion figure for hypochondria's annual cost to the health care system comes from Jerome Groopman, "Sick with Worry," *The New Yorker*, August 11, 2003, which is an excellent piece on the basics of health anxiety.

The full stories of Henrietta Darwin and Sara Teasdale can be found in Baur, *Hypochondria*.

On the Microsoft "cyberchondria" study, see John Markoff, "Microsoft Examines Causes of 'Cyberchondria,'" *New York Times*, November 25, 2008.

For an entertaining and humorous—albeit highly opinionated—memoir-style overview of anxiety, see Patricia Pearson, *A Brief History of Anxiety: Yours and Mine* (New York: Bloomsbury, 2008).

The Pac-Man-style video game study can be found in Dean Mobbs et al., "When Fear Is Near: Threat Imminence Elicits Prefrontal-Periaqueductal Gray Shifts in Humans," *Science* 317, no. 5841 (August 24, 2007).

For more on Kendler's studies on the heritability of anxiety, see Kenneth Kendler et al., "The Genetic Epidemiology of Irrational Fears and Phobias in Men," *Archives of General Psychiatry* 58, no. 3 (March 2001), and Kenneth Kendler et al., "The Etiology of Phobias," *Archives of General Psychiatry* 59, no. 3 (March 2002).

Michelle Craske discusses the differences in anxiety and fear expression between the sexes exhaustively in her book *Origins of Phobias and Anxiety Disorders: Why More Women Than Men?* (Oxford: Pergamon, 2003).

The Robert Leahy quotation about uncertainty is from his helpful book *The Worry Cure: Seven Steps to Stop Worry from Stopping You* (New York: Three Rivers Press, 2006).

Tom Borkovec and his collaborators have written scores of papers and articles on worry, but a useful summary of their research can be found in the chapter titled "Worry: Unwanted Cognitive Activity That Controls Unwanted Somatic Experience," by Borkovec and Lizabeth Roemer, in Daniel Wegner and James Pennebaker, editors, *Handbook of Mental Control* (Englewood Cliffs, N.J.: Prentice Hall, 1993).

Daniel Wegner discusses his "white bear" research in great detail in his book *White Bears and Other Unwanted Thoughts: Suppression, Obsession, and the Psychology of Mental Control* (New York: Guilford Press, 1994). His study on the paradoxical results of trying hard to relax is in Daniel Wegner et al., "Ironic Effects of Trying to Relax Under Stress," *Behaviour Research and Therapy* 35, no. 1 (January 1997).

For more on Bill Tancer's research into fear-based search queries, see his book *Click: What Millions of People Are Doing Online and Why It Matters* (New York: Hyperion, 2008).

The post-9/11 research on how fear shaped Americans' perceptions of the risk of being injured in a terrorist attack is in Jennifer S. Lerner et al., "Effects of Fear and Anger on Perceived Risks of Terrorism: A National Field Experiment," *Psychological Science* 14, no. 2 (March 2003). Somewhat hilariously, while survey respondents rated others' chances of being hurt by terrorism in the next year at 48 percent, they claimed that their own chance of being injured was just 21 percent. Should we be insulted that these people seem to think of us as brainless cattle with a death wish?

The William James quotation comes from his *The Principles of Psychology*, volume 2 (New York: Henry Holt, 1918).

Jean Twenge's rather dismal findings on the psyches of America's youth can be found in her book *Generation Me: Why Today's Young Americans Are More Confident, Assertive, Entitled—and More Miserable Than Ever Before* (New York: Free Press, 2007).

For a scientific overview of fear extinction, see Francisco Sotres-Bayon et al., "Brain Mechanisms of Fear Extinction: Historical Perspectives on the Contribution of Prefrontal Cortex," *Biological Psychiatry* 60, no. 4 (August 15, 2006).

More on David Barlow's interoceptive exposure treatment, as well as an exhaustive survey of anxiety research and history, can be found in his *Anxiety and Its Disorders: The Nature and Treatment of Anxiety and Panic*, 2nd edition (New York: Guilford Press, 2004).

For an excellent scientific summary of Davis's d-cycloserine research, see Michael Davis et al., "Facilitation of Extinction of Conditioned Fear by D-Cycloserine," *Current Directions in Psychological Science* 14, no. 4 (August 2005).

The best, most succinct, and most factual life of the Buddha I've found is Karen Armstrong's *Buddha* (New York: Penguin, 2001). Among the many, many books on mindfulness and meditation, Bhante Henepola Gunaratana's *Mindfulness in Plain English*, updated and expanded edition (Somerville, Mass.: Wisdom Publications, 2002), is a true gem.

On the mindful-noting neuroscience study, see Matthew Lieberman et al., "Putting Feelings into Words: Affect Labeling Disrupts Amygdala Activity in Response to Affective Stimuli," *Psychological Science* 18, no. 5 (May 2007).

For the best overview of acceptance and commitment therapy, see Steven Hayes's self-help book *Get out of Your Mind and into Your Life: The New Acceptance and Commitment Therapy* (Oakland, Calif.: New Harbinger Publications, 2005). The summary of studies on ACT's efficacy can be found in Steven C. Hayes et al., "Acceptance and Commitment Therapy: Model, Processes and Outcomes," *Behaviour Research and Therapy* 44, no. 1 (January 2006).

For a nice example of Danny Forster's fear of heights in action, see a clip of his Shanghai experience at www.youtube.com/watch?v=pH5p2CxOrhY.

CHAPTER THREE: *The Zen of Shock Trauma: Stress, Strain, and Coping with Chaos*

For more on Baltimore's Shock Trauma unit and its ever-increasing lifesaving prowess, see Michael Rosenwald, "Yesterday, They Would Have Died," *Popular Science*, October 2003.

One could practically fill a bookstore with guides to stress, but the standout in the field is Robert Sapolsky's *Why Zebras Don't Get Ulcers*, 3rd edition (New York: Holt Paperbacks, 2004).

The study on New Yorkers and heart attack risk is in Nicholas Christenfeld et al., "Exposure to New York City as a Risk Factor for Heart Attack Mortality," *Psychosomatic Medicine* 61, no. 6 (November/December 1999).

An excellent discussion of the psychological effects of the London Blitz bombings can be found in S. J. Rachman, *Fear and Courage*, 2nd edition (New York: W. H. Freeman, 1990). Similarly, there is a very useful discussion of the psychological effects of shellfire in World War II in William Ian Miller, *The Mystery of Courage* (Cambridge, Mass.: Harvard University Press, 2000).

For more on the research into combat stress, see Ben Shephard, *A War of Nerves: Soldiers and Psychiatrists in the Twentieth Century* (Cambridge, Mass.: Harvard University Press, 2001), as well as Miller's *The Mystery of Courage* and Bourke's *Fear: A Cultural History*.

On the link between sitting in traffic and heart attack risk, see Annette Peters et al., "Exposure to Traffic and the Onset of Myocardial Infarction," *New England Journal of Medicine* 351, no. 17 (October 21, 2004), which chronicles the German team's earlier research.

Two good overviews of stress at work can be found in Anahad O'Connor, "Cracking Under the Pressure? It's Just the Opposite, for Some," *New York Times*, September 10, 2004, and "Working Long Hours? Take a Massage Break, Courtesy of Your Boss," *New York Times*, September 7, 2004.

For an intriguing look at the new air traffic training techniques, see Matthew L. Wald, "For Air Traffic Trainees, Games with a Serious Purpose," *New York Times*, October 8, 2008.

Frederick Lanceley distills his knowledge of hostage negotiation tactics in his book *On-Scene Guide for Crisis Negotiators*, 2nd edition (Boca Raton, Fla.: CRC Press, 2003).

On Morgan's Special Forces research, see Andy Morgan and Maj. Gary Hazlett, "Assessment of Humans Experiencing Uncontrollable Stress: The SERE Course," *Special Warfare*, Summer 2000.

A useful discussion of Farley's risk-taking research can be found in Richard Restak, *Poe's Heart and the Mountain Climber: Exploring the Effects of Anxiety on Our Brains and Our Culture* (New York: Harmony Books, 2004).

For much more on Maddi's resilience research, see Salvatore R. Maddi and Deborah M. Khoshaba, *Resilience at Work: How to Succeed No Matter What Life Throws at You* (New York: AMACOM, 2005).

A helpful introduction to Gary Klein's decision-making research can be found in Bill Breen, "What's Your Intuition?," *Fast Company*, August 2000. Klein's own book on the topic is *Sources of Power* (Cambridge, Mass.: MIT Press, 1998).

An excellent magazine feature on Ericsson's studies at the Human Performance Laboratory is John Cloud, "The Science of Experience," *Time*, February 28, 2008. Ericsson's own infinitely cited early research paper on practice and elite performance is K. Anders Ericsson et al., "The Role of Deliberate Practice in the Acquisition of Expert Performance," *Psychological Review* 100, no. 3 (July 1993). Deakin and Cobley's figure skating research can be found in their paper "An Examination of the Practice Environments in Figure Skating and Volleyball: A Search for Deliberate Practice," in Janet L. Starkes and K. Anders Ericsson, editors, *Expert Performance in Sports: Advances in Research on Sport Expertise* (Champaign, Ill.: Human Kinetics, 2003).

For more on the Vietnam POWs and humor, see Linda D. Henman, "Humor as a Coping Mechanism: Lessons from POWs," *Humor— International Journal of Humor Research* 14, no. 1 (June 2001).

Rod Martin's comprehensive book on humor is *The Psychology of Humor: An Integrative Approach* (San Diego, Calif.: Academic Press, 2006).

CHAPTER FOUR: *Think Fast: Cognition Under Pressure, and How to Improve It*

Ogi Ogas cataloged his *Millionaire* performance and preparations in two pieces: "Who Wants to Be a Cognitive Science Millionaire?," *Seed Magazine*, November 9, 2006, and "The Decider," *Boston Globe*, November 11, 2006.

A transcript of Yerkes and Dodson's classic paper on arousal level and performance, "The Relation of Strength of Stimulus to Rapidity of Habit-Formation," can be found at psychclassics.yorku.ca/Yerkes/Law.

Antonio Damasio discusses case studies demonstrating the effects of amygdala damage on cognition in his book *Descartes' Error: Emotion, Reason, and the Human Brain* (New York: Putnam, 1994).

For an excellent exploration of the psychological aftermath of the *War of the Worlds* broadcast, see the episode of WNYC's radio program *RadioLab* titled "War of the Worlds," which first aired on November 3, 2008. (It's available at the WNYC website, www.wnyc.org, and on iTunes.) Also, see Dozier's *Fear Itself.*

A wonderful overview of Slovic's risk assessment research can be found in Paul Slovic et al., "Risk as Analysis and Risk as Feelings: Some Thoughts About Affect, Reason, Risk, and Rationality," *Risk Analysis* 24, no. 2 (April 2004). It's also available at www.decisionresearch.org/pdf/dr502.pdf. For a more layman-oriented feature on risk analysis, see Jeffrey Kluger et al., "Why We Worry About the Things We Shouldn't…and Ignore the Things We Should," *Time*, December 4, 2006.

On driving deaths after the September 11 attacks, see Garrick Blalock et al., "Driving Fatalities After 9/11: A Hidden Cost of Terrorism," *Applied Economics* 41, no. 14 (June 2009).

Steele and Aronson's landmark paper on stereotype threat is Claude M. Steele et al., "Stereotype Threat and the Intellectual Test Performance of African Americans," *Journal of Personality and Social Psychology* 69, no. 5 (November 1995). The study on working memory and math performance is in Sian L. Beilock and Thomas H. Carr, "When High-Powered People Fail: Working Memory and 'Choking Under Pressure' in Math," *Psychological Science* 16, no. 2 (February 2005).

On "helmet fire" and airline crashes, see James E. Driskell and Eduardo Salas, editors, *Stress and Human Performance* (Mahwah, N.J.: Lawrence Erlbaum, 1996).

For more on United Airlines Flight 232, see the excellent episode of Errol Morris's documentary series *First Person* entitled "Leaving the Earth," which tells the tale through Dennis Fitch, the flight instructor who helped in the cockpit.

Chapter Five: *Before the Madding Crowd: Decoding the Mysteries of Performance Anxiety and Stage Fright*

Baumeister's friendly audience research can be found in Jennifer L. Butler and Roy F. Baumeister, "The Trouble with Friendly Faces: Skilled Performance with a Supportive Audience," *Journal of Personality and Social Psychology* 75, no. 5 (November 1998). For much more on blowing it in front of audiences, see Roy F. Baumeister, Todd F. Heatherton, and Dianne M. Tice, *Losing Control: How and Why People Fail at Self-Regulation* (San Diego, Calif.: Academic Press, 1994).

On Blascovich's pet-audience study, see Karen M. Allen et al., "Presence of Human Friends and Pet Dogs as Moderators of Autonomic Responses to Stress in Women," *Journal of Personality and Social Psychology* 61, no. 4 (October 1991).

For a brisk overview of stage fright, see John Lahr, "Petrified," *The New Yorker*, August 28, 2006.

More on Bill Russell and his pregame nerves can be found in John Taylor, *The Rivalry: Bill Russell, Wilt Chamberlain, and the Golden Age of Basketball* (New York: Random House, 2005). The Ella Fitzgerald anecdote comes from Stuart Nicholson, *Ella Fitzgerald: A Biography of the First Lady of Jazz* (New York: Da Capo Press, 1993).

I drew my information on Laurence Olivier's famous stage fright primarily from his autobiography, *Confessions of an Actor: An Autobiography of Laurence Olivier* (New York: Simon & Schuster, 1982), but also from Terry Coleman, *Olivier* (New York: Henry Holt, 2005), and Lahr's "Petrified."

Paul Salmon and Robert Meyer's outstanding and concrete book on managing stage fright is *Notes from the Green Room: Coping with Stress and Anxiety in Musical Performance* (San Francisco: Jossey-Bass Publishers, 1998).

For more on the brains of elite marksmen, see Amy J. Haufler et al., "Neuro-Cognitive Activity During a Self-Paced Visuospatial Task: Comparative EEG Profiles in Marksmen and Novice Shooters," *Biological Psy-*

chology 53, nos. 2–3 (July 2000). A similar analysis of cognition in elite golfers can be found in John Milton et al., "The Mind of Expert Motor Performance Is Cool and Focused," *NeuroImage* 35, no. 2 (April 2007).

For more on beta-blocker use in classical music performance, see Blair Tindall, "Better Playing Through Chemistry," *New York Times*, October 17, 2004.

The illusion of transparency research is in Kenneth Savitsky and Thomas Gilovich, "The Illusion of Transparency and the Alleviation of Speech Anxiety," *Journal of Experimental Social Psychology* 39, no. 6 (November 2003).

On the research into cognitive reappraisal and its effect on fear expression, see Elizabeth Phelps, "Neural Circuitry Underlying the Regulation of Conditioned Fear and Its Relation to Extinction," *Neuron* 59, no. 5 (September 2008).

On facilitative versus debilitative views of anxiety in competitive swimmers, see Graham Jones et al., "Intensity and Interpretation of Anxiety Symptoms in Elite and Non-Elite Sports Performers," *Personality and Individual Differences* 17, no. 5 (November 1994).

The Gandhi anecdotes and quotations can be found in his book *An Autobiography: The Story of My Experiments with Truth* (Boston: Beacon Press, 1993).

CHAPTER SIX: *The Clutch Paradox: Why Athletes Excel — or Choke — Under Fire*

The definitive contemporary piece on Steve Blass's struggles is Roger Angell's "Gone for Good" (originally titled "Down the Drain"), *The New Yorker*, June 23, 1975, which has often been called one of the best sports features ever written, and for very good reason. (Blass even told me he felt honored to be a part of Angell's piece, which is not a sentiment one frequently hears about the media these days.) Also excellent is Pat Jordan's "Pitcher in Search of a Pitch," *Sports Illustrated*, April 15, 1974.

Gray and Beilock offer a very useful overview of athletic choking in their essay "Why Do Athletes Choke Under Pressure?," which can be found in Gershon Tenenbaum and Robert C. Eklund, editors, *Handbook of Sport Psychology*, 3rd edition (New York: John Wiley and Sons, 2007).

An invaluable resource for information on the psychology of baseball is Mike Stadler's aptly titled *The Psychology of Baseball: Inside the Mental Game of the Major League Player* (New York: Gotham Books, 2007).

The legendary anecdote of Larry Bird at the three-point contest comes from Dan Shaughnessy, "Bird Gets His Way Thanks to His Will," *Sports Illustrated*, February 12, 1986.

On Gray's choking research with college baseball players, see Rob Gray, "Attending to the Execution of a Complex Sensorimotor Skill: Expertise Differences, Choking, and Slumps," *Journal of Experimental Psychology: Applied* 10, no. 1 (2004).

Baumeister's classic choking experiment is in his article "Choking Under Pressure: Self-Consciousness and Paradoxical Effects of Incentives on Skillful Performance," *Journal of Personality and Social Psychology* 46, no. 3 (March 1984). On verbal suggestion and choking at the free throw line, see Larry M. Leith, "Choking in Sports: Are We Our Own Worst Enemies?," *International Journal of Sport Psychology* 19, no. 1 (1988).

The rock climbing study is in J. R. Pijpers et al., "Anxiety-Induced Changes in Movement Behaviour During the Execution of a Complex Whole-Body Task," *Quarterly Journal of Experimental Psychology* 58, no. 3 (April 2005). The soccer dribbling study is in Sian L. Beilock et al., "When Paying Attention Becomes Counterproductive: Impact of Divided Versus Skill-Focused Attention on Novice and Experienced Performance of Sensorimotor Skills," *Journal of Experimental Psychology: Applied* 8, no. 1 (2002).

On Wegner's "ironic effects," see Daniel Wegner et al., "The Putt and the Pendulum: Ironic Effects of the Mental Control of Action," *Psychological Science* 9, no. 3 (1998).

The famous John Candy story can be found in Anne Ryan, "Poise: Vinatieri Gets His Kicks in the Clutch," *USA Today*, January 29, 2004.

The quotation on obsessive-compulsive athletes comes from Larry Stone, "The Art of Baseball: A Tradition of Superstition," *Seattle Times*, September 25, 2005.

Wade Boggs's incredible game day routine is detailed in Peter Gammons, "All in the Preparation," *Sports Illustrated*, April 14, 1986.

Gmelch's excellent paper on "baseball magic" is "Superstition and Ritual in American Baseball," *Elysian Fields Quarterly* 11, no. 3 (1992). An updated edition can also be found online.

For more on Vinatieri, see Michael Lewis's great piece "The Kick Is Up and It's...a Career Killer," *New York Times*, October 28, 2007.

On athletic visualization, see Shane M. Murphy, "Imagery Interventions in Sport," *Medicine and Science in Sports and Exercise* 26, no. 4 (1994).

For more on Dennis Eckersley, see Steve Wulf, "The Paintmaster," *Sports Illustrated*, August 24, 1992; Scott Ostler, "Eckersley's Saving Grace: Worrying Keeps A's Relief Ace from Falling out of Groove," *San Francisco Chronicle*, October 6, 1992; and Ron Kroichick, "Eck's Long Haul to Fame," *San Francisco Chronicle*, July 11, 2004.

The case against clutch performance ability is a pet cause of many baseball statisticians, and the argument is well made in Joe Sheehan, "The Concept of

'Clutch,'" BaseballProspectus.com, March 10, 2004. Also, see Dave Sheinin, "The 'Clutch' Conundrum," *Washington Post*, April 6, 2008, and Tom Verducci, "Does Clutch Hitting Truly Exist?," *Sports Illustrated*, April 5, 2004. The Grabiner study can be found at www.baseball1.com/bb-data/grabiner/fullclutch.html.

On the illusive "hot hand," see Thomas Gilovich et al., "The Hot Hand in Basketball: On the Misperception of Random Sequences," *Cognitive Psychology* 17, no. 3 (July 1985).

For more on Matthews's stress and driving research, see Gerald Matthews et al., "Driver Stress and Performance on a Driving Simulator," *Human Factors* 40, no. 1 (1998).

An excellent study on the links between confidence and performance in baseball can be found in Ronald Smith and Donald Christensen, "Psychological Skills as Predictors of Performance and Survival in Professional Baseball," *Journal of Sport & Exercise Psychology* 17, no. 4 (December 1995).

The story of Larry Bird's late-game performance in Seattle can be found at www.nba.com/features/birds_coaches.html.

CHAPTER SEVEN: *Mayday, Mayday!: How We React, Think, and Survive When Our Lives Are on the Line*

My account of Cafferata's icy night on Toktong Pass owes much to Larry Smith, *Beyond Glory: Medal of Honor Heroes in Their Own Words* (New York: W. W. Norton, 2003), and to Peter Collier, *Medal of Honor: Portraits of Valor Beyond the Call of Duty* (New York: Artisan, 2006). For more on the Chosin campaign, see Michael Kernan, "Chosin Survivors and the 78-mile Nightmare," *Washington Post*, December 1, 1984, and J. Robert Moskin, "Chosin," *American Heritage*, November 1, 2000.

Abner Small's words about Crane and *The Red Badge of Courage* come from his book *The Road to Richmond: The Civil War Letters of Major Abner R. Small of the 16th Maine Volunteers* (New York: Fordham University Press, 2000).

An excellent source on combat performance is Shephard's *A War of Nerves*. Also invaluable are Lt. Col. Dave Grossman (with Loren W. Christensen), *On Combat: The Psychology and Physiology of Deadly Conflict in War and in Peace* (Millstadt, Ill.: Warrior Science Publications, 2008), and Bourke's *Fear: A Cultural History*.

S. L. A. Marshall's classic combat research book is *Men Against Fire: The Problem of Battle Command in Future War* (Gloucester, Mass.: Peter Smith, 1978).

On the hand-holding study, see Richard J. Davidson et al., "Lending a Hand: Social Regulation of the Neural Response to Threat," *Psychological Science* 17, no. 12 (December 2006).

For much, much more on Alex Horton's experiences in Iraq, see his exceptional blog Army of Dude at armyofdude.blogspot.com.

For the details on Lionel Wigram, I am primarily indebted to Shephard's *A War of Nerves*.

The bible of information on human behavior in survival situations is John Leach's strangely hard to find *Survival Psychology* (New York: New York University Press, 1994). Also excellent is Amanda Ripley's *The Unthinkable: Who Survives When Disaster Strikes—and Why* (New York: Crown, 2008), which includes extensive sections on the civilian response to the 9/11 attacks and the Beverly Hills Supper Club inferno.

On the misconception about mass panic in emergencies, see Lee Clarke, "Panic: Myth or Reality?" *Contexts*, Fall 2002.

For more on the Tenerife airport disaster, see Amanda Ripley, "How to Get Out Alive," *Time*, May 2, 2005.

On freezing in a crisis, see John Leach, "Why People 'Freeze' in an Emergency: Temporal and Cognitive Constraints on Survival Responses," *Aviation, Space, and Environmental Medicine* 75, no. 6 (June 2004).

Two useful sources for Alexis Artwohl's research on perceptual distortions are her papers "Perceptual and Memory Distortion During Officer-Involved Shootings," *FBI Law Enforcement Bulletin*, October 2002, and "No Recall of Weapons Discharge," *Law Enforcement Executive Forum* 3, no. 2 (2003).

Eagleman's time perception experiment can be found in Chess Stetson, Matthew P. Fiesta, and David M. Eagleman, "Does Time Really Slow Down During a Frightening Event?," *PLoS One* 2, no. 12 (December 2007).

On Morgan's scuba diving death research, see William P. Morgan, "Anxiety and Panic in Recreational Scuba Divers," *Sports Medicine (NZ)* 20, no. 6 (December 1995). Also, Bernie Chowdhury chronicles the underwater lifestyle in his book *The Last Dive: A Father and Son's Fatal Descent into the Ocean's Depths* (New York: HarperCollins, 2000).

Grossman provides quips, anecdotes, and limitless information about training in *On Combat*.

For more on wilderness survival, see Peter Kummerfeldt's short book *Surviving a Wilderness Emergency* (Colorado Springs, Colo.: OutdoorSafe Press, 2006), or Laurence Gonzales's *Deep Survival: Who Lives, Who Dies, and Why* (New York: W. W. Norton, 2004).

CONCLUSION: *How to Be Afraid*

Daniel Stockwell has shared his story publicly on only a few occasions, but for those who are curious, his wife told it from her viewpoint in Merry

Stockwell (with Cal Fussman), "You Do It Quiet," *Esquire*, November 1998. The tales of more Carnegie Medal winners can be found in the commission's book *A Century of Heroes;* request a copy at www.carnegiehero.org/book_request.php.

The classic work on the original Mercury program astronauts is Tom Wolfe's fascinating *The Right Stuff* (New York: Bantam, 1983). Two other useful sources for the Gordon Cooper spaceflight story are Cooper's autobiography (with Bruce Henderson), *Leap of Faith: An Astronaut's Journey into the Unknown* (New York: HarperCollins, 2000), and Loyd S. Swenson, Jr., James M. Grimwood, and Charles C. Alexander, *This New Ocean: A History of Project Mercury*, a NASA special publication, which can be found online at history.nasa.gov/SP-4201/toc.htm.

The Ron Howard quotation comes from Charles S. Warn, "Ron Howard Weightless Again over *Apollo 13*'s DGA Win," *Director's Guild of America Magazine*, www.dga.org/news/mag_archives/v21-2/howard.html. The Buzz Aldrin quotation is in his op-ed "Fear and Flying," *New York Times*, February 3, 2003.

For more on Carol Dweck's intelligence research, see her piece "Beliefs That Make Smart People Dumb," in Robert J. Sternberg, editor, *Why Smart People Can Be So Stupid* (New Haven, Conn.: Yale University Press, 2002).

Index

academic achievers, 149–50
acceptance and commitment therapy
    (ACT), 83–85, 191, 279, 282
acrophobia, 78, 85–90
adrenaline
    fight-or-flight response and, 25, 26, 28,
        29, 61, 134
    neuropeptide Y and, 112
    physical tasks and, 137
    propanolol and, 51, 178, 181–82
    TV game shows and, 158, 160
agoraphobia, 44n
air traffic control, 105–8
Aldrin, Buzz, 271–72
amygdala
    cognition and, 36–41, 136, 157–58
    danger and, 36, 38–39, 41, 50
    disabling/damaging, 39, 137
    exposure to fears and, 88, 275–77
    fear compared to anxiety and, 58–59
    fight, flight, or freeze response and,
        36, 237
    independent sensory connections of,
        36–39, 281
    memory and, 42–46, 49, 247
    as neural command center of fear, 30, 65
    NMDA and, 77–78
    prefrontal cortex and, 74–75, 81–82,
        126, 185
    primal fears and, 44–45
    relationship with fear and, 35–41, 52
    representations of fearsome stimuli and,
        40–41, 74

    role in fear, 35–36
    spontaneous recovery of fears and, 72–74
    stage fright and, 165, 167, 168
    unwanted fears and, 50, 51
Angell, Roger, 194
antianxiety medications, 11, 77
antidepressants, 11, 77
anxiety
    attention and, 205–10
    author's experience with, 14–15, 282
    avoidance and, 13, 70–72, 223, 275–76
    breathing and, 66n, 158, 254, 273
    cognition and, 58, 63–65, 137
    definition of, 33
    distorted thinking and, 67–68, 81
    fear compared to, 58–59
    genetic heritage and, 59–61
    as health issue, 11
    heroism and, 264–65
    hypochondria and, 54–55, 57
    mindfulness and, 78–85
    modern life and, 10–12, 52, 68–69
    optimal anxiety levels, 136–37, 181
    prefrontal cortex and, 58, 70
    stress compared to, 94–95
    student test-takers and, 148–51, 274
    uncertainty and lack of control and,
        61–63, 89–90, 98–105, 276–77, 282
    worry and, 13, 63–67, 275
    worsened by bad coping strategies, 13,
        174–77
    See also choking under pressure; fear;
        stage fright; stress

"'It's The People' is a core value at Intuit. This means we are high per-forming teams focused on achievement. We put customers at the heart of everything we do. And we innovate to drive growth and continuously improve all that we do. We take risks and grow by learning from our mistakes. So, how do we reinforce these expectations, behaviors, and desired outcomes? Recognition is the key... recognition is an integral part of a successful culture that's a strategic lever for ongoing perfor-mance feedback. This is why CEOs care deeply about the culture of their companies, and Mosley and Irvine offer an excellent approach for fostering and nurturing such a winning culture of recognition."

—Jim Grenier
Vice President, Rewards, Workplace, & HR Global Shared Services
Intuit, Inc.

"This is a great 'how to' book in transforming your culture. Eric and Derek have captured the fundamental reasons on why organizations need to rethink recognition platforms. As a driver of this change in my organization, I can attest to the positive impact the reworking of the program has had. The book is full of 'aha' moments and information is presented in a way that shows you that change is not difficult. A must read."

—Thomas G. Aurelio
Vice President, Human Resources
Symantec

"Recognizing mastery, communicating purpose, encouraging autonomy, Eric and Derek offer practical guidance on how to create a company culture that feeds our true motivators."

— DANIEL H. PINK, AUTHOR OF
DRIVE: THE SURPRISING TRUTH ABOUT WHAT MOTIVATES US

WINNING
WITH A CULTURE OF
RECOGNITION

by ERIC MOSLEY and DEREK IRVINE

Recognition Strategies at the World's Most Admired Companies

✳ ✳ ✳
Acknowledgments

Before we begin, working within our own culture of recognition, now is the time to call out the much appreciated and valuable contributions made by many people to this book.

First off our thanks goes to Andrea Dumont and Doug Hardy for helping us with the production of this book. We laughed a lot, had some fun, but during tough decision times, which would have sunk less experienced folks, their total professionalism helped carry us through smoothly. We also want to say a big thanks to Lynette Silva for her substantial research over many years now, and her keen editorial eye. All three have been pivotal contributors to this book in various ways.

A huge call-out to our customers. Without them and their enthusiastic support for everything we in Globoforce have been doing over the past decade, none of this would have been possible. It's particularly satisfying for us when our customers follow our recommended strategies and enjoy tremendous success; they then go on to become great advocates and champions for everything about strategic employee recognition. We have also learned a great deal from our customers. We appreciate greatly your continued enthusiasm for what is now a revolution in the thinking about culture management. A particular call out to those who participated in this book: Avnet, Dow Chemical, Fairmont Hotels & Resorts, Intuit and Symantec.

And finally, to each and every employee of Globoforce, your actions everyday are what make our company special, what brings to life our thinking. We never forget that without our collective focus on our own values and on what we need to do to succeed, none of this would be possible. This book is dedicated to you.

✳ ✳ ✳
Table of Contents

✱ ✱ ✱
Introduction

The Monthly Executive Meeting, Carametric Systems, Inc.

"I still don't get it," said Tom. "What does it take to change the Carametric culture? We spend a fortune hiring the right people. We spend time and money talking up our products. But the tech blogs and reviewers down at CNET say we've lost our technical edge." The CEO tapped his Blackberry. "Now our best customers are telling me the same thing in personal emails to me complaining that Carametric's products are behind the curve. They're asking us what happened to the old spirit of innovation?"

David, the chief technology officer, said, "We've been over this before. Our customers want reliable, not flashy. Their systems depend on stability, and their investors care about continuity."

Tom replied, "So you're telling me we can't be both reliable and innovative? There are plenty of companies whose cultures do both. Look at Google. Look at GE. Look at Intuit. What's our problem?"

David shrugged. Everyone remained silent until Laura, the chief people officer and head of human resources, spoke.

"We brought this problem on ourselves. We put so much emphasis on reliability across our tech team that we ended up killing innovation and risk-taking."

Laura passed copies of a one-page document around; it contained

Employee Values Distribution Graph

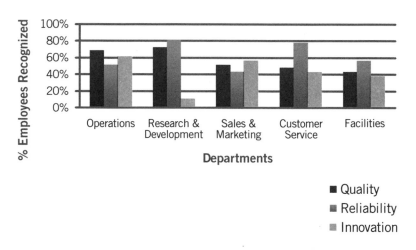

a simple bar graph. "As you know, we've been tracking the cultural shift we want to make. This is the instrument I've been showing you—the employee values distribution chart. Do you notice something different from last month's graph?"

"Looks the same to me," said Tom, "except you've got a new bar here." He looked around the table. "The orange bars on the right are new. What do they stand for?"

Laura replied, "Three months ago, when we began our discussion about innovation, I added a category for employee behaviors that managers should recognize. We've done training and we've done communication. We've seen a rise in creativity and new ideas all over the company, from customer service to facilities management.

"But I'm sorry to say, David," she added, "that the managers in the research and development department are registering the lowest employee recognition rate for innovation. Your guys and gals in R&D don't get recognized for innovation; that means that even if they're *interested in the next big thing, they're not bringing that behavior to work, or worse yet, they're not being encouraged to take risks. The facts are there."*

David pursed his lips and silently studied the chart.

Tom spoke. "Okay, that's clear enough. Innovation's happening everywhere but where it matters most. David, stay after this meeting with Laura and me and we'll figure out how to fix the situation."

Tom concluded, "We're betting the company on this new initiative, ladies and gentlemen. We'd better get innovation right."

✳

Organizational culture is the most powerful force in business, and yet it is one of the most neglected (and misunderstood) attributes of any organization. Every organization has a culture, whether it is the culture the leadership wants or the one that has come to exist through inertia and management neglect. Businesses are societies, and society and culture are inseparable.

In evaluating the culture of our clients' workplaces, the questions we ask are:

* Is your organization's culture deliberately managed or left to develop by accident?
* Is your culture aligned to and supportive of your company values?
* Does your culture contribute every day to your strategic goals or hinder them?
* And most importantly—can you prove your answers to these questions are correct?

Understanding your company's culture and shaping it deliberately, based on your values, is critical to achieving your strategic objectives. It is critical to gaining competitive advantage. We at

Globoforce are in the business of managing culture—yes, managing it. To say that corporate culture cannot be managed scientifically, with rigorous and authentic processes, is a myth with damaging consequences. An organization's culture can be learned, encouraged, ingrained, and applied to every business process, in many forms and across many different parts of the organization. Applied correctly, culture management through recognition is one of the most powerful, effective and, most critically, positive ways to drive the success of your organization as measured by improvements in operating margins, income, and customer satisfaction.

This book serves as a guide for business leaders to harness the power of that culture to achieve the company mission. Think of the hallmarks of your culture—your company values. These statements of desired behaviors and actions suggest and—when followed properly—reinforce your culture.

✳ Executive Insight

"I really focus on the values and the standards of the organization. What are the expected behaviors? How do we want to treat each other? How do we want to act? What do we want to do about transparency? How can we have a safe environment where we really know what's going on?"

—ALAN R. MULALLY,
President and Chief Executive Officer of Ford Motor Company [i]

But how do you make your values real to every employee so they are meaningful in how employees approach their everyday tasks? Strategic recognition, implemented correctly, is the most powerful tool we know to manage, inspire, encourage, and measure demonstration of the values you have chosen for your organization.

In this book we'll teach you how simple appreciation—strategically applied—achieves these goals. The book is organized into two main parts:

Part I—A Recognition Framework

We'll explain a three-level, progressive model of culture management through recognition; we'll build on foundations of management science, research, and our experience with leading corporations like GE and Intuit to propose a new model, which we call *strategic recognition*.

Chapter 1: Recognition and Culture—This chapter explains how culture drives values through an organization, and why most efforts at manipulating culture through recognition fail. We'll describe the three-stage evolution of recognition and the role of culture in several well-known organizations.

Chapter 2: How Recognition Works—This chapter discusses how a method of continual positive reinforcement works with individual employees and organization-wide to create, encourage and manage a chosen culture. We'll show how and why recognition enables employees and organizations to reach their fullest potential.

Chapter 3: Strategic Recognition—This chapter examines the breakthrough concept that moves recognition from a "nice to have," to an invaluable practice fully integrated with global management systems.

Part II—Implementing Strategic Recognition

Chapter 4: Build a Strategic Recognition Practice—This chapter outlines a ten-stage plan for implementing strategic recognition in your organization, with step-by-step strategies and key practices that will improve your culture and results.

This book represents the best thinking from the best-performing organizations on culture, employee engagement, and the simple human truths underlying successful management practices. The best practices presented here can help you and your organization create a culture of appreciation that drives performance, profits, and pride.

✳ ✳ ✳

***

PART I

A Recognition Framework

✳ ✳ ✳
Chapter 1
RECOGNITION AND CULTURE

When Tom said that he and the executive team are "betting the company" on innovation, he was referring to a huge change in its product line. Starting next year, Carametric will join the cloud computing movement by introducing a software-as-a-service (SaaS) product to customers. Carametric's competitors already offer SaaS to their customers because they are more innovative. They hire, manage, and reward innovation. Tom realizes that he has to make his firm more innovative at every level. Innovation has to become part of Carametric's culture.

*Laura's bar chart, called a **values distribution graph**, showed everyone at the meeting that innovation was underway in various departments, but that the critical research and development group, David's group, lagged in this area.*

Carametric's new SaaS-based product requires new learning and new effort from employees in research and development, marketing, sales, and customer care—that is, everyone in the company. Current products or customers cannot be neglected, so this initiative will require extra effort, ingenuity, and innovation from people who weren't hired based on their creativity. So the strategic question is—how do Tom and his leadership team foster innovation as a daily habit throughout the company?

*The answer is simple—by constantly **recognizing, celebrating, and reinforcing** innovation. When the majority of employees notice the right behaviors that foster innovation, call them out, praise them, reward them, and put out extra effort to reinforce those behaviors, then the com-*

pany's employees have gone beyond **liking** *the new idea to* **living** *the new idea, thereby making innovation part of the company's culture.*

As a strategic move, creating a new company value will take time. Tom will hire more innovative and creative people in the future, along with a few specialists in SaaS, but typically it is not new employees who create new values. In fact, entrenched cultures resist change, especially from outsiders, which is one reason that David is having trouble bringing innovation to his team.

Alternatively, Tom could create an elite minority within the company, one that emphasizes innovation. That minority will, however, battle uphill against cultural inertia at the implementation phase. Contemplating that "culture war," Tom decided there had to be a better way. He wasn't going to change Carametric's culture by himself or with a cadre of elites. Everyone at the company had to participate in the movement to innovation, or it would never become part of the culture. Tom asked Laura's advice, and she had a ready answer.

"You have to make innovation a habit," she said. "Recognize and reward innovation everywhere. Get people talking about it. Give them a chance to figure out what innovation means, and when they do, celebrate it at every turn."

"That sounds okay in theory," said Tom. "But how are you going to know it's working? We don't have five years to make this change."

"We'll follow it, Tom. We can watch this change as closely as you watch the revenue numbers," Laura replied. "And with the right data, we'll manage the process. It can be done."

✳

How can a spirit of innovation become self-sustaining? At what point in all this effort can Tom and his senior management stop hectoring, cajoling, and making the case for change to employees? How will the managers and their staff take on innovation as a habit, as part of their daily routine? Tom has to reach out beyond a small, elite group and change the culture of Carametric to become more innovative from top to bottom.

Now imagine a Carametric in which acts of innovation and creativity are recognized and rewarded. Imagine managers rewarding their staff for innovative acts and ideas, and peers recognizing colleagues for the same things. Imagine front-line employees publicly recognizing innovative managers and executives. Imagine every individual at every level being deputized to spot and reward innovation.

Further, imagine that Laura and Tom could track, quantify, and measure this activity every day, then correlate its impact on performance and achievement of the desired change. With the right tools, Laura and Tom can spot where innovation is happening and where it's lagging, encouraging the former and improving the latter.

What would happen to Carametric? It would *change itself* to become more innovative. Innovation would become part of its identity because the people who make up the organization would receive constant, genuine, authentic, and meaningful reinforcement of this new value. Peers would not only notice each others' efforts, but celebrate them as well. Managers could be evaluated on how effectively they inspire innovation. Innovation would be at the top of everyone's mind.

In this changed culture, a major initiative like SaaS is a natural outgrowth of the living value, innovation—a value that is now directed toward achieving a business goal.

An organization's culture is so much more than a slogan or poster. Culture is nothing less than the aggregate of tens of thousands of interactions every day. Leaders of great companies reinforce their values by rewarding and celebrating the behaviors that express those values.

The power of corporate culture was expressed a few years ago in the influential work of John P. Kotter and James L. Heskett. In *Corporate Culture and Performance* they wrote, "[Corporate cultures] can enable a group to take rapid and coordinated action against a competitor or for a customer. They can also lead intelligent people to walk, in concert, off a cliff."

Is your organization's culture leading your people to success or leading them off a cliff?

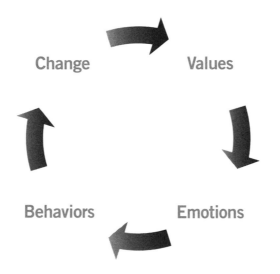

What is the energy driving this cycle? Emotions. When a manager recognizes an employee's behavior, personally and sincerely, both feel proud, gratified, and happy. There's a human connection that transcends the immediate culture to create a shared bond. The power of this bond is stronger than you might think; indeed, it's the power that holds together great organizational cultures. Shared values, shared emotions, shared connections—these make organizations as much as they make civilizations.

✳ Executive Insight

"Culture is a slow-growing tree. In the beginning it needs protection. But after a couple of decades the culture will be stronger than you are. You need to work with it, not against it."

"Culture is a powerful but fragile thing. If you burn down the culture tree, it takes a long time to grow another one."

—WALLY BOCK,

Three Star Leadership [ii]

Creating a culture means choosing a limited number of values that define the company as surely as its products or logo, and then encouraging expression of those values in everyday behavior. No single set of values defines culture; greatness lies in authenticity. At Nike, for example, the culture includes keen competitive spirit. At Apple, designers will not put a power cord on a device that doesn't look incredible and is seamlessly aligned with the overall design of the product, because fabulous design is a top priority of the Apple culture. Some firms, like Ryanair and Walmart, thrive on driving down costs, while others, like BMW and Rolex, focus on premium-priced engineering. If you care more about driving cost out of the system, you belong at Ryanair or Walmart, not BMW or Rolex.

Furthermore, as any psychologist will tell you, positive reinforcement works better than negative reinforcement to foster a particular corporate culture. This is just as true for organizations as it is for individuals.

✳ The Power of Positive Reinforcement

Even in tough-minded cultures, positive reinforcement is a powerful driver of culture. U.S. Marine training might be strenuous and even abusive, but that initiation process is not the culture. The stress of Marine boot camp serves as much to *identify* Marines as to train them. Once established as a Marine, an individual experiences profound recognition on a daily basis—reinforced by the mottos, the uniform, the unit cohesion, the intense group loyalty. Marines display recognition for their service and sacrifice on their uniforms in the form of medals, ribbons, and rank insignia. All these inspire pride and internal reward. Marine culture is intensely about recognition. Watch two retired Marines talking—twenty years after their service ended, they'll still call each other "Marine."

Let's return to that big product change at Carametric. Imagine at this moment, all over the world, employees are trying innovations of all kinds. Sometimes the "new way of doing things" is directly related to the new initiative, and sometimes it's just a small improvement in an old process. People in customer service are building a more customized database of replies to customer questions. People in operations are refreshing their Six Sigma certificates. And every time this happens, someone notices. There are little pieces of goodwill happening between people. People are recognizing their colleagues, leaders, and employees for discretionary effort, for living the new company value: Innovate.

The spirit of innovation is becoming part of the company

culture, and all of the positive reinforcement is inspiring greater engagement and greater emotional involvement in the staff. Senior leaders notice more discretionary work. People are listening to one another. Formerly timid employees are speaking up. People who used to feel cynical about their chance of making a difference now know that they matter and push themselves and others to do better. People know what's expected of them and help to create a spirit of innovation everywhere.

As positive emotions are continuously linked to new behaviors, those new behaviors are sought out and multiply quickly, leading to profound culture change. This is the power of recognizing, celebrating, and reinforcing the right behaviors.

The root causes of this positive culture change comes down to simple human emotion. Consider these two stories within the Carametric culture.

*

John works for the tech division of Carametric. He has been a loyal and dedicated employee for more than a quarter century. In his annual performance reviews, the only form of feedback or recognition he receives, John always earns high ratings and is commended for his excellent work ethic and the quality of his work. But John has never once received— from his boss or anyone on his teams—so much as a casual pat on the back or a "Good job! Thanks for all your work."

One day John happens to be at the office's printer when he sees a document with his name on it. Taking it back to his desk to read, John discovers that the engineers with whom he worked on his last project completed a detailed evaluation form about his work and contribution to the team. John realizes that this is a standard form completed after every project, but he has never seen or even heard of it before.

Reading through the comments, John is initially pleased to see the engineers reporting only positive news, praising his insightful work and suggestions that actually earned his company extra money on the contract for early program delivery. But after several weeks of waiting, John does not hear anything from his supervisor about the form, the compliments in it, or the money his efforts saved the company.

The entire experience serves only to de-motivate John. It's as if his excellent work isn't acknowledged by his supervisor, even when others praise his results. John begins to notice that he's not the only one whose work goes unrewarded, unacknowledged, and unappreciated. Saying "Thank you" just isn't part of his group's culture. Soon, John is feeling alienated to the point where he considers leaving the company. When a friend emails John to ask if he should apply for a developer's position at Carametric, John writes back, "You don't want to work here."

Sarah's story is much more encouraging. Sarah works as an administrator in Carametric's finance group, and the group has embraced Carametric's formal recognition program. Sarah has been repeatedly recognized by her direct supervisor as well as her peers for her efforts to ensure that employee benefits are processed quickly and accurately. She's paid particular attention to Carametric's tuition assistance program, a benefit that CEO Tom began to encourage all employees to take advantage of in order to increase their skills. Fellow employees have commended Sarah to her manager through the recognition program, acknowledging how Sarah's efforts have enabled them to return to school and improve their performance.

In no small part because of this frequent recognition of her work, Sarah is happy in her role with no desire to leave Carametric. She is motivated and believes that Carametric is a great place to work. She's said so on her Facebook page, and last month Sarah starred in a short video on the recruiting page of Carametric's website, saying how proud she is to work in a department that recognizes her unique contributions.

✳

While these two, real-life stories of recognition in the workplace may seem to convey simple truths on the surface, the lessons run far more deeply. The experiences of both John and Sarah illustrate the differences in culture between their two departments, and what behaviors those subtle differences reinforce. John will have no desire to go the extra mile in a culture that demands much but rewards little. Sarah will do anything she can to promote the success of her department and the organization as a whole.

These examples illustrate two exciting, and demanding, realities in the modern workplace:

1. Employees expect more than a paycheck;
2. Authenticity is golden.

Employees expect more than a paycheck. Have you, like John, ever worked in an organization in which management talks about what the culture is, and the employees silently think, "Yeah, yeah, you speak about this value, but your behavior says otherwise." This cynicism in the face of management is epidemic. It's one reason we laugh at Dilbert cartoons. It's one reason for public rage at how irresponsible financial institutions worldwide caused the Great Recession of 2008–2009. On the other hand, executives who show authenticity by facing the brutal truth in a forthright manner or who give every employee, no matter their position, a stake in the company's growth (as when Google's on-site massage therapist became a millionaire through the company's stock options) engender legitimacy. These acts of authenticity, of living values with integrity, even to the point of sacrificing some reward, give enormous power to their authority. Employees in survey after survey say that they want to work for an authentic, open, and appreciative culture.

Authenticity is golden. In the age of the Internet, when employees anonymously share information about the inner workings and culture of a company, the truth will come out. The bright side of this golden coin is that when a company's leadership effectively rewards and recognizes corporate values, the company will be celebrated in cyberspace by its own employees. On Facebook, on Twitter, on LinkedIn, in email, and on job message boards, employees will confirm the company's legitimacy to each other and to the world. That kind of advertising can't be bought, and that kind of criticism can't be countered with a press release about "company values." In sustaining a culture, you have to walk the talk.

Organizational Culture

Herb Kelleher, the legendary cofounder and chairman of Southwest Airlines, believed that "Culture is what you do when people aren't looking." It's the result of how employees behave when they step away from the power relationships in an organization and operate purely out of their own values. When those values are also shared with the organization, culture is nourished.

✱ **Research Insight**

"Fifty-one percent of Gen Y workers are prepared to accept a lower wage or a lesser role if their work contributes to something more important or meaningful."

—KELLY SERVICES,
Kelly Global Workforce Index [iii]

It's easy enough to see this in companies where certain values are paramount. Take JetBlue Airways. Anne Rhoades, the creator of JetBlue's People Team, recalls how the airline's interest in safety caused her to hire a mechanic no other company would touch:

"I asked one of the first technicians we hired, 'Give me an example of a time when safety was an issue but it was a very difficult thing to own up to because you knew that your superior did not agree.' He said, 'I graduated at the top of my class and went to work for a New York airline, a job I had wanted all my life. About sixty days into the job, we had a plane going overseas that I did not feel was perfectly safe. There were two or three items that I thought we should fix before I signed off. I went to my supervisor, and he said that I needed to sign off without the fixes.'"

Anne continues: "He didn't sign off and got fired. No other airline would touch him. We hired him because he put his career on the line for a safety issue. That shows incredible integrity. He's now a supervisor at JetBlue."

Since culture is manifest in behaviors in ways that are both as dramatic as this once-in-a-career moment and as mundane as regular day-to-day behaviors, we have to ask, what guides those behaviors? What can managers do to establish and maintain their culture?

Certainly they can lead by example and inspiration, but not everyone can imitate a great leader (even if they try). Management needs a tool that helps all employees learn for themselves how to "live the culture."

Recognition is that tool, and there are three levels of value to it.

Individual Recognition

Individual recognition is at the center of everything we write about in this book because it's where the connection happens between a

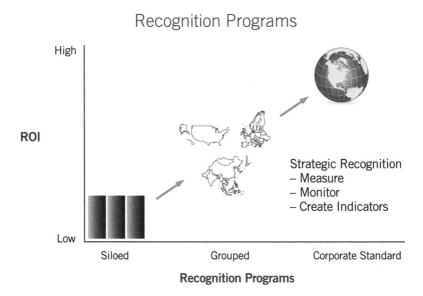

manager and an employee. The manager notices extra effort or good performance and recognizes it with an award and a personal message connecting the behavior to a value that is part of the company culture. There are two connections, actually—the association between behavior and values, and the human connection between manager and employee. (Incidentally, when we say "manager and employee" we mean everyone from line managers to top executives. Recognition works at all levels.)

"Catch them doing the right thing" is the slogan for this kind of recognition. Proponents focus on the recognition moment—the positive interaction between manager and employee. They encourage managers to find many little ways to celebrate employee performance so that positive reinforcement becomes a habit.

We call this *individual recognition* because it is really a one-to-one connection. Most of us have seen managers who just had a knack for recognizing and motivating their people. It's positive effort, and it motivates employees to become more enthusiastic about their jobs. On its surface, the chief appeal of individual recognition is its appearance of being easy and fun. People like to give and receive a pat on the back. (In fact, it's a sad comment on most corporate cultures that implementing individual recognition is an improvement on what was happening before.)

An unmanaged individual-recognition program works best in a small business, which is typically driven by the personality and leadership of the owner. Mallory's Appliance Repair, with ten employees, is a small society in which Mallory can celebrate and recognize good work frequently. General Electric, with 350,000 employees, is a different story. The vast majority of employees are part of many layers of management or are thousands of miles removed from executives. Managers with diverse personalities, cultural expectations, and leadership abilities cannot be expected to behave in the same way or convey the same critical messages. An individual-recognition program that relies on creative and spontaneous impulse is out of place in the complex, data-dependent multinational corporation. Why? Because you can't possibly deliver this spontaneous behavior across thousands of individuals through hope, encouragement, or cajoling alone. Managers don't manage anything else on hope, so why recognition? In such a workplace, systems need to be in place to encourage the right behavior in the appropriate scale.

For a large organization to implement an individual-recognition program in a sustainable way, it can't just depend on the enlightenment of every manager and his or her ability (or willingness) to find 1,000 ways to give trinkets to employees. Furthermore, a large organization risks complete chaos in its recognition efforts if it can't provide a scalable way for all managers to recognize their people.

Enterprise Recognition

In the last few decades, an entire industry grew up to create more formal recognition programs for organizations of about 100 employees or more. Recognition companies provided simple guidelines and basic training for managers who might not otherwise know the most effective ways to recognize their people. They offered large catalogs of awards and prizes that managers could give to employees.

This service, which we call *enterprise recognition*, was the next step in the evolution of a recognition practice. It encouraged managers to participate in recognition. It gave some amount of quality control to the people in charge of recognition, typically someone in the human resources department. It made recognition efforts standard and scalable across big organizations, just as a purchasing department can standardize and scale the buying of office supplies.

What happens when you take a good idea like individual recognition and try to scale it across the enterprise? Typically, you invest in the supporting structures of a recognition program. You create or buy different tools to scale training, communication, and the physical delivery of awards. You make the program global. You make it sustainable in terms of making sure that everybody has access to the right budgets and it's happening all over the enterprise, in different divisions and in different languages. You've added another layer of value to the delivery of individual recognition and ensured that it's delivered company-wide.

However, delivering a company-wide, universally adopted recognition program requires a break from the past in terms of typical program design. Old-fashioned recognition programs are notoriously hard to scale:

* Old-fashioned recognition programs demand creativity from some managers who just don't have it. (Its acolytes try to overcome this problem by promoting lists with names like "365 ways to reward employees.")

* Old-fashioned recognition programs, as typically practiced, focus awards around previously chosen objects, such as a lapel pin, a USB flash drive, a cupcake, a desk set, a plaque with the company logo. The value of the gift is subjective and variable according to the recipient's taste and the giver's choice.

* Old-fashioned recognition programs' focus on these objects complicates implementation across cultures in large global companies. The programs tend to be U.S.-focused and thus fail to account for cultural differences that would demonstrate respect and understanding of employees' local culture.

* Old-fashioned recognition programs are difficult to manage in terms of directing costs to best use, tracking the business benefits of the practice, and following up. Focused on the individual manager and the recognition moment, these programs generally rely on ad hoc reporting and subjective impressions.

* Old-fashioned recognition programs, because they require such creativity and depend on manager buy-in, are hard to enforce and typically have low penetration rates.

Enterprise recognition is a thoroughly modern approach. However, it is not effortless to launch or easy to sustain in a modern organization. Certainly if you design the program incorrectly at the start, you'll have great difficulty in scaling it up so that it reaches 80 to 90 percent of the workforce. Standardized solutions create friction in a large, complex department. There are logistical issues around ordering and shipping merchandise. People often don't participate because they have never heard of the program or they don't like the awards. In a global organization, there are language, currency, and tax complications.

Enterprise recognition falls short of recognition's true potential because it is layered onto a culture in the same manner as a benefit program. It stays in the human resources silo—a positive step forward, to be sure, but not answerable like other disciplines to management practices of measurement against goals. That represents a significant missed opportunity, one that can be captured by the final evolution in recognition, which we call *strategic recognition.*

Strategic Recognition

Strategic recognition is the practice of integrating recognition with other management practices, taking recognition beyond the human resources silo and leveraging its power to shape behavior at all levels of the organization. (When we say *power*, we mean recognition's unique ability to help employees manage themselves, as opposed to just obeying directions from the "powers that be.") When individual recognition moments across the enterprise are recorded, analyzed, and understood, recognition becomes as potent a management tool as financial- or program-management practices.

Strategic recognition adds the ultimate layer of value, which is *culture management.* Strategic recognition is linked to strategic goals such as engagement, employee satisfaction, or culture change. But also, because you have those tools, you get to then use strategic recognition to manage the culture. In other words, you can emphasize a single value that you feel doesn't have the traction you need to meet your strategic objectives.

Stopping the effort at individual or even enterprise recognition is shortsighted, like focusing on one quarter's sales to the exclusion of long-term sales analysis or product development. Your sales strategy doesn't focus exclusively on the moment a sales transaction closes (at least, it shouldn't). Strategic sales practices track and analyze customer relationships, planning, product development, marketing intelligence, follow-through, and growing skills of sales staff. In the same way, strategic culture management needs a long-term set of practices that make the moment of recognition reflect a larger drive to fulfill company strategy.

Strategic recognition takes its place with the other "hard" management science practices. It has measurable processes. It is fully integrated into strategic planning and global resource management. It removes barriers to success. Self-sustaining, strategic recognition can enhance and define organizational culture, bring certain values to the surface, and drive a culture in which behaviors reflect organizational values and contribute to company success.

✳ Research Insight

"For most companies, recognition is an underutilized asset, one that you can—and should—set on the right track. Your recognition programs telegraph what you value and what you want to happen; recognition is how your employees perceive what they are supposed to do. So if you're unsure of whether your message—or strategic plan, or shift in culture—is getting through, a well-run recognition program can tell you."

—CAROL PLETCHER,
The Conference Board Review [iv]

Strategic recognition aligns company culture to geographic, national, and even demographic cultures. The company's most important values are understood by everyone: young Europeans and older Asians, jocks in the financial planning department, hipster designers in marketing, and minivan-driving soccer parents in the call center. Strategic recognition becomes so much more than the relationship between manager and employee—it becomes the affirmation of belonging to the society we call a corporation. Its goal is not to continually add new incentives but to become a self-sustaining set of desired behaviors—to create an organizational culture.

The Heroic Leader

Worldwide, organizations celebrate the heroic leader—the man or woman whose vision and will create (and presumably enforce) a particular culture. This celebration parallels the rising fascination with celebrity itself: *People* and other magazines celebrate fame even as they create it. Business media picked up this trend, glorifying leaders such as Lee Iacocca of Chrysler, Bill Gates of Microsoft, Steve Jobs of Apple, Lou Gerstner of IBM, Richard Branson of Virgin, and Herb Kelleher of Southwest Airlines. Even those whose success turned out to be a sham (Enron's Kenneth Lay, "Chainsaw Al" Dunlap of Sunbeam, the financier Bernard Madoff) were hailed in their time as cultural engineers of rare vision. Each had definite ideas about how their company cultures should operate. Each was the antithesis of the boring or nearly anonymous corporate leadership at other companies in their industries as well as the "organization men" of the 1950s and 1960s. And each made a compelling and colorful figure.

Lesser leaders bought into the notion that one person created the unique cultures that led to their organizations' success. They tried to imitate the leader's singular, heroic position as the one indispensable person around whose vision all activity took place.

The imitators rarely succeed because they credit the individual and his vision alone. However, this heroic leader image rarely builds lasting organizations.

Many CEOs hope that the senior team will extend that cult of personality by passing on what the boss says or translating the boss's values into their individual styles as they manage their departments. Soon, multiple vice presidents are working within their own spheres of influence.

In practice, the cultural norms and imperatives that the CEO feels are important become diluted further. Managers pursue different business drivers, different imperatives, different problems and opportunities. Social hierarchies complicate the picture: If the company has a go-go sales culture, then the sales representatives are the royalty. If it's a product culture, the product managers are the princes and princesses of the realm.

In the business classic *Good to Great: Why Some Companies Make the Leap...and Others Don't*, Jim Collins describes the "level 5 leader" as a frequently colorless public figure, yet one focused on many of the social mechanisms of an organization. His analysis shows the drawbacks of a heroic leader, including the obvious fact that an organization dependent on one person for its success gets in trouble when that person leaves, retires, or dies. Great leaders, says Collins, live more often out of the public view, but build behavior-based values like "facing the bad news" and a "culture of discipline" into their organizations' DNA. Great leaders enable the organization to build culture itself. They hire people who will promote the right cultural values. They envision a *social architecture* that supports the culture they want.

Social Architecture

Social architecture is to culture what a foundation, beams, and joists are to a building. Social architecture is the scaffolding of a company: communication, traditions, authority, privileges, and "ways of doing things." It includes behavior cues like how people dress and how they talk to one another. It includes how excellence is recognized and rewarded because it's a way of talking about the implementation of culture.

Social architecture takes a value like "determination" and translates it into situational behaviors like, "No matter what, we will never give up on a sale." It is the framework of communication, positive and negative reinforcement, public and private knowledge, and cultural cues that determine how the company will operate. It includes the hierarchy of authority and reward, the transparency of information, and even the manners and traditions of the company.

Social architecture exists because no manager can be everywhere, on every phone call, standing beside every employee whenever they're doing anything. It's the set of behavioral norms that define a culture—"what you do when nobody is looking."

Jack Welch led GE through enormous changes—the Work Out program for breaking down bureaucracy and hierarchy and Six Sigma processes are just two he oversaw—and his vision, uncompromising standards, and astute use of media aided his success. GE was and is a vast, worldwide organization with hundreds of thousands of employees in scores of countries. He couldn't meet with every employee to persuade them to perform to his standards in his way. He needed a methodology and a structure that would nurture the values he deemed most important. He needed employees to act according to GE values even when nobody was looking. So he posited a set of principles that defined how work would be measured, evaluated, and judged.

GE's managers were evaluated on how thoroughly they applied

Welch's principles to their divisions and departments. GE had been a social entity and a successful company for 100 years before Welch took the reins; that social architecture was available to Welch to drive his initiatives throughout the organization. Adapting along the way, he improved communication and impressed every employee with his determination to fight for his values. To cite one example, Welch's dictum to "fire the bottom 10 percent" based on objective performance criteria promoted hard-headed management and fairness simultaneously. The rating and shedding of poor performers yearly became part of GE's social architecture, one that promoted achievement and a sense of being the best.

✳ Executive Insight

> *"The middle 70 percent are managed differently. This group of people is enormously valuable to any company; you simply cannot function without their skills, energy, and commitment. After all, they are the majority of your employees. But everyone in the middle 70 needs to be motivated, and made to feel as if they truly belong. You do not want to lose the vast majority of your middle 70—you want to improve them."*
>
> —JACK WELCH,
> *Winning* [v]

Social architecture doesn't require a public figure like Welch to be enormously effective. Some of the most successful companies in the world have had a succession of "quiet" CEOs. Johnson & Johnson, for example, expresses its values in "Our Credo," a statement that describes its responsibilities to doctors, nurses, patients, families, employees, communities, and, finally, shareholders. Outlined are specific behavioral guidelines ("Compensation must be fair and adequate, and working conditions clean, orderly, and safe") that are flexible enough to apply across countries, businesses, and cultures. The details in Our Credo are numerous, but the key concept is responsibility to others, a sense that each person's work has an impact on many others, and that individuals within Johnson & Johnson are accountable for that impact.

Johnson & Johnson's success illustrates one advantage of scale: The global reinforcement of values endows them with great power. When a large environment is aligned along just a few values, there is little ambiguity. A hundred signals a day promote their adoption. (Any decent manager or line worker can determine whether a workplace is "clean, orderly, and safe.") Scale adds power, however, only if a large majority of employees express the organization's values in their behavior.

At the other end of the size spectrum, the start-up organization also benefits from a deliberate social architecture. Start-ups classically begin with a few people, a vision, and an obsessive focus on just one or two central ideas. That focus is critical when a company is small and the CEO can promote an idea face-to-face with 10 or 50 or 100 employees. But daily interaction with employees for a start-up executive becomes impossible as the company grows. It's just a fact of life. There will be people the CEO won't see on a daily basis, so the cult of personality wanes as the company grows to 50 employees or more.

Zappos.com is a great example of successfully translating the values of a hero CEO deep within an organization as it experiences explosive growth. A key value for Zappos.com is its nonnegotiable, obsessive, 24/7 devotion to customer service. The company's CEO,

Tony Hsieh, built the company around it. (Motto: "At Zappos. com, Customer Service Is Everything. In Fact, It's the Entire Company.") Early on, Tony and his executive team could promote these values by power of example, asking, "How will this affect the customer?" at every opportunity, and by rewarding and recognizing workers who shared the company's obsession with customer satisfaction. As the company grew, Tony and his team could hire likeminded managers who demonstrated their passion for customer satisfaction in their actions. Today the company has sales of more than $1 billion, and each new employee is hired on the basis of their values—putting the customer first—and publicly or privately honored for any demonstration of that value. When Jeff Bezos, Amazon.com's CEO, announced the purchase of Zappos.com in July 2009, he credited the company's obsession with customer service for his decision. Bezos said that such a company made him "weak in the knees"—to the tune of around $900 million—and added that he wasn't going to change a thing.

✱ Executive Insight

"What he [Hsieh] really cares about is making Zappos.com's employees and customers feel really, really good because he has decided that his entire business revolves around one thing: happiness. Everything at Zappos.com serves that end."

"Zappos.com's 1,300 employees talk about the place with a religious fervor. The phrase core values can prompt emotional soliloquies, and the CEO is held with a regard typically afforded rock stars and cult leaders."

—INC. MAGAZINE [vi]

Three components of social architecture deserve special mention here. They are shared values, engaged employees, and united execution. Shared values, employee engagement, and united execution create a high-performance culture. Strategic recognition is the link connecting all three.

Shared Values

Management teams of most companies have spent countless hours concisely defining their company's values and honing their company's mission into an ambition that inspires employees to achieve strategic goals. In reality, these values rarely move beyond the engraved plaque hanging on the wall. For a company's values to have an impact on employee behavior and performance, they must be understood in the same way by all employees regardless of position, division, or geographic location.

Beyond being understood, values should differentiate a company from its competitors. The company's unique value set needs to be promoted, rewarded, and propagated.

Shared values are taught, retaught, and honored when recognition draws attention to specific behaviors tied to a company value. Of course, the individual being recognized is reminded of the values demonstrated. If the recognition is public or requires approval the behaviors and values are doubly reinforced. If all recognition within a set time is shared in a team meeting, then entire teams will be reminded of the company's values. In large, globally distributed companies this is virtually the only way to make the company values come alive for every employee.

To achieve this level of common understanding, managers must clearly and consistently communicate the organization's values. However, this task can be diluted not only by a company's scale but also by a varied and diverse workforce (especially in global companies). Furthermore, the task of communicating is complicated by variability in communication skills among managers and individual manager's perceptions of the relative worth of certain values. To manage is to choose among multiple options, and business situations inevitably cause a manager to choose in the moment between, for example, customer satisfaction and greater efficiency. Deeply ingrained values point the way to resolve these conflicting options.

Engaged Employees

Employee engagement is the HR buzzword of the decade. Engaged employees are enthusiastic and involved; they are personally invested in better performance. Engagement means doing more than the job requires; it also implies that the urge to do more comes from within, as opposed to "just following orders." Discretionary effort is at the heart of engagement.

The connection between engagement and higher performance is obvious, and while there are many ways to inspire engagement (and even more to kill it) all involve communicating and rewarding desired behaviors based on defined values.

Executives desire it, consultancies specialize in it, and numerous good studies assert its positive effect on the bottom line. To give just one example, a study by Towers Perrin in 2003 documented that "companies with higher employee engagement outperform those with lower employee engagement, relative to industry benchmarks" like revenue growth, cost of goods sold, and customer focus. Among 40 multinational companies, Towers Perrin found higher operating margins and net profit margins in the firms with more engaged employees. In a book describing the insights of those studies [vii], Towers Perrin's Julie Gebauer wrote, "We consistently found that organizations and managers get the best from employees when they do five things well: know them, grow them, inspire them, involve them, and reward them. When these five principles are at the core of the work experience, there's no doubt that employees consistently give value-adding discretionary effort—and that directly impacts the organization's financial results."

As to management methods for fostering engagement, a Northwestern University study found, "There are three climate factors that positively influence engagement, including training, autonomy, and personal power. ...All of these factors are related to feelings on the part of the employee regarding personal value, respect, and freedom. Thus, engagement is largely driven by the employee feeling that the organization values his or her contribution."

✳ Executive Insight

"Employee engagement is an experience to be lived, not a problem to be solved."

—DAVID ZINGER,

Global Employee-Engagement Expert

Done right, recognition communicates the right behaviors, rewards and reinforces those behaviors, and gives a sense that the organization values employee contributions. Done right, recognition is management's positive way of creating engagement.

Engaged employees are important contributors to the company's culture and continually reinforce values that support the company mission as well as the bottom line. At its most powerful, recognition continuously propagates and reinforces desired behaviors throughout the company.

✳ Executive Insight

"Changing the culture to reward the desired behavior is critical to success. Make heroes of day-to-day deliverers, not those who make the biggest splash. You reward people on how they treat the customer, how they make decisions, how they simplify the business. ...And, crucially, all of this has to be done in the spirit of open communication and respect. ...If [people are] uncertain and they don't feel respected, the change will never stick. Celebrating success, recognizing achievement, and making people feel good about the business were important tools for sustaining momentum. Importantly, it's as much—if not more—about the recognition of your peers than it is about financial rewards."

—FIONA MACLEOD,

Business Unit Leader, Convenience Retail Americas, BP [viii]

How important is engagement to the individual? The number-one reason people leave their current employer is the feeling they don't count, that their work was not recognized. "It wasn't valued. My contributions weren't appreciated," is the common complaint. This leads to disaffection and alienation—the psychic opposite of engagement.

United Execution

All organizations are simply people united by common goals; employees need to work together to achieve those goals. Even with shared values and great enthusiasm, a team that is not united in its efforts is dysfunctional. Once individuals unite as a team, functioning together to achieve departmental, divisional, and company goals, they can reinforce both values and engagement. Mutual dependence develops trust, encourages learning, and fosters the sense of belonging to something greater than oneself.

Human nature does not always lend itself to putting the team ahead of personal goals, however, and the nature of employment in fast-moving organizations and national cultures (short job tenure, emphasis on resume building and individual achievement) can further discourage unity. (Ironically, many of the fastest-moving organizations, such as high-tech firms, are heavily dependent on unity of execution because their work is far too complex for individuals to complete.)

Management and, indeed, the culture itself must encourage teamwork by directly rewarding it and by demonstrating its importance. What do we mean when we say the culture encourages teamwork? It's as simple as a director instructing her managers, "Spot all the good behavior you can, because I can't be everywhere." In a consistent culture, this is manifest in a thousand person-to-person moments, connecting the value of teamwork to specific behaviors in a direct, personal, and individual recognition moment.

By making the company's values come alive for all participants and rewarding employee behaviors that reinforce those values, appropriate and consistent recognition across teams and countries inspires the workforce to achieve great things. United execution is recognized as a method that furthers both individual and organizational goals. Recognition of teamwork provides psychological reward to the individual, but only if she or he is invested in the group.

Hiring to Fit the Company Culture

Deliberate management decisions, not happenstance, create a high-performance culture. For employees, this begins with the hiring process.

Organizations that value their culture recruit people who are likely to behave in the way that expresses that culture. In an article in the *New York Times*, reporter Claire Cain Miller described the hiring system at Zappos.com:

"Employees go through two sets of interviews, one about qualifications and one to see if they fit the culture. All employees work in the call center for a month. After a week, they can take the pay they have earned plus $2,000 and leave. Only about 3 percent do [according to CEO Tony Hsieh], but it weeds out those who are not committed to the company. Half of annual performance reviews assess how the employee fits in to the company culture." [ix]

Fairmont Hotels & Resorts, another splendid example of effective hiring, spends a lot of time and money during its hiring process to identify and confirm a candidate's dedication to outstanding service. Fairmont looks for employees who want to delight hotel guests. This requires hard-to-measure qualities like empathy, creativity, and spontaneity. Matt Smith, Fairmont's executive director of learning and development, gives this example of a successful hire:

"We had guests staying with us in one of our resorts in the Rocky Mountains in a room that had a big stone fireplace. As they left the room to go swimming, one of their children said, 'Mom, I can't believe there's a fireplace. Do you think they'll know how much I love marshmallows? Could we roast marshmallows?' Well, a room attendant—the person cleaning the room—happened to overhear that conversation. When the family returned, they found a basket of marshmallows, graham crackers, and chocolates, all to make s'mores around the fire. On the basket was a little handwritten note from the employee saying, 'Because we know how much you like marshmallows.'"

"You can't engineer that kind of creativity," adds Matt. "You can't write a manual that says, 'If you ever have a kid and a fireplace, send marshmallows.' What you can do is recognize the magic that spontaneous, creative service creates and keep hiring people who want to make that magic. And you can manage a continuous process of turning those unpredictable moments into organizational culture through appreciation, communication, and celebration of those acts."

✳ Tips for Praising and Appreciating Employees Successfully

Give specific praise that goes far beyond a generic "Great job!" to make recognition truly meaningful. With specific praise, you tell the recipients *what* they did, *how* that behavior/effort reflected the company values, and *why* it was important to the team/department/company or contributed to achieving strategic objectives.

Praise actions that you want to see repeated. By giving employees such specific recognition, you clearly communicate what is important and encourage them to repeat those actions in the future. For employees to want to repeat such desired behaviors, however, you must...

Make the praise and recognition authentic. Don't fall into the compliment sandwich trap by saying things like, "Great job on that task, but you forgot this one critical step. I know you'll get it next time, since you are so conscientious!" This is a confusing message to employees. Did they really do a good job if an important step was missed? Offer constructive criticism, which is also desired by employees, separately from praise for work well done.

✳ ✳ ✳
Chapter 2
HOW RECOGNITION WORKS

Tom, David, and Laura remained in the conference room after the other executives adjourned.

"David, I'll get right to the point," said Tom. "Innovation's not happening in R&D. Not enough, not fast enough. You're supporting the move to SaaS. Do you need new people on your team?"

"No," said David. "My people are always looking out for the next big thing. They're just focused on first things first, which is reliability. I know they can innovate if they need to. I've seen it."

"When you saw it, what did you do?" asked Laura.

"I let the R&D managers decide if a new feature should be added to the product," David replied, his voice quickening impatiently. "You might know that in software development you have to test code before you release it..."

Laura smiled. "I mean, what did you do to show the employee that what he or she was doing was important? Did you recognize the fact that something new was being created?"

David waved her away. "My people know when they're appreciated..."

Tom cut him off. "No, they don't, Dave. That's exactly what the values distribution graph tells us." He said to Laura, "All the managers in R&D get trained in employee recognition. How come it's not happening?"

"It starts at the top," Laura said. To David, she suggested, "I think you set the tone for the entire group, David. We need some very public and

clear demonstrations that innovation is what you want and what you'll reward."

✱

The transformative power of rewarding behavior to drive values deep into an organization relies on clear and consistent communication at all levels and in every location of a company. No tool is more powerful for achieving this than strategic recognition.

Managers will ask, "What's the point of recognition? Employees do their jobs and I pay them. Why should I do anything more?" To answer this question, we need to look at what drives workplace behavior and the extent to which managers can inspire increased engagement and thus greater performance.

Managers who rely on pay alone to get desired performance will meet some limited success. Employees who are just looking for the security of a paycheck will generally show up and perform the tasks in their job description. This is a recipe for mediocrity. In fact, this point of view demonstrates an obsolete view of management itself, for the manager must continually prompt, cajole, and direct behaviors.

The usefulness of recognition to the manager lies in making the right behaviors and attitudes self-sustaining. To understand how recognition accomplishes this, consider the psychological effect of different rewards.

Employees think of a salary or hourly wage as the minimum contract between them and their employer. A business friend we know is fond of saying, "Whatever I pay someone, it won't be enough in six months. They'll get accustomed to the paycheck arriving (imagine their reaction if it didn't), but they'll stop relating it to the growing, ever-changing challenges of their jobs. Some other reward has to feel as fresh and immediate as today's big project."

The social architecture of every business bears complex interactions in which factors like power, prestige, friendship, affinity, hostility, prejudice (negative and positive), security, confidence, trust, and faith inspire action. In fact, it's not just the *reality* of these factors, but their *perception*, that matters a great deal to whether a person feels engaged in his or her work.

Into this welter of emotions wades the manager, trying to inspire great performance and discourage mediocrity with a few tools. Compensation is a critical tool, but in most jobs it is the one the manager can affect least. In brief, a paycheck is the minimum requirement of the working relationship—no work, no pay.

Psychic Income

Human beings have a fundamental need for social acceptance, increased self-esteem, and self-realization. [x] In a business setting, these needs can never be met by cash compensation, which organizational psychologist Fred Hertzberg found could only prevent people from being dissatisfied. Salary is what we call tangible income—vitally important, but related to material needs as well as status or power.

Study after study shows that *nonmonetary rewards* are the key to improved performance. These rewards, which we call *psychic income*, are cost-effective as well. They are more flexible, affordable, and immediate than salary.

Psychic income is the provision of social acceptance, social esteem (leading to self-esteem), and self-actualization. Paid in the "currency" of recognition, psychic income is intangible but no less real than material income.

This is reminiscent of leadership guru Stephen R. Covey's metaphor of the "Emotional Bank Account." [xi] Covey's model portrays acts like courtesy, respect, and honesty as deposits in the account, and discourtesy, disrespect, and dishonesty as withdrawals. The balance in Covey's Emotional Bank Account is the amount of trust in a relationship.

✳ Research Insight

More than half (53 percent) of those surveyed say that their boss is dishonest, and the same amount say that their boss is unfair. Others described bosses as impatient (58 percent), disloyal (66 percent), and lacking motivational skills (76 percent). These numbers are staggering, given that nearly all (89 percent) say that the employee/boss relationship is one of the most important links to job satisfaction.

—ADECCO GROUP NORTH AMERICA [xii]

So it is with psychic income. Managers and executives pay out psychic income to employees with acts of respect, esteem, dignity, and high regard. They reduce psychic income with acts of disrespect, humiliation, disinterest, and low regard. The balance between these is the amount of psychic wealth accumulated at work.

Organizational psychologist Fred Hertzberg identified only one management tool—recognition—that could result in employee satisfaction because only recognition feeds psychic income needs. And you don't have to wait until payday to make a deposit.

Incentives vs. Recognition

An executive may respond, "What about incentives for performance? When my sales staff exceeds their goals, they get bonus pay. When my managers perform exceptionally, they get profit-sharing (cash or stock options). We hold incentive competitions in which the top five performers get a trip to Cancún. Isn't that incentive enough?"

Yes, incentives can be effective management tools, but recognition is qualitatively different. It's not a question of whether management needs one or the other, because recognition inspires a different set of psychic rewards than incentives. The table below contrasts the qualities of incentives versus recognition.

Incentives	Recognition
Objective targets	Subjective behavior
Known reward *(no surprise)*	Unknown reward *(surprise)*
Known frequency	Unknown frequency
Infrequent *(e.g., annual bonuses)*	Frequent *(every hour, every day)*
Tangible reward primary	Intangible reward primary
Intangible reward secondary	Tangible reward secondary
Numbers-based	Values-based
Focused on elite few	Focused on many

For example, incentives are earned based on objective targets. The reward is based on milestones agreed in advance, almost always in terms of financial performance. For a sales executive, there's a clear numerical connection between closing the sale and getting the incentive. Most employees don't have that clear of a connection; their daily performance has a minimum requirement but no direct incentive to perform better. Recognition, on the other hand, rewards behavior based on values, culture, and other less easily quantified *but no less important factors.*

Like salary, incentives operate as a relationship between the employee and the organization, again based on financial performance. Recognition operates more as a direct relationship between the employee and his or her manager, and this is a critical difference. According to one aphorism, "People join organizations but leave managers." Recognition fosters a positive relationship with the boss.

With incentives, the number is the primary reward—it's all about keeping score. (Exceed the quota by 10 percent—get X reward. Exceed it by 20 percent—get 2X reward.) With recognition, the primary rewards are prestige, pride, satisfaction, and other psychological rewards that can far exceed the actual monetary value of the recognition given.

Recognition rewards behavior in real time or soon after. Incentive requires more time between the action and the reward.

Note that recognition is not restricted to honoring intangible values. Some of the most effective recognition programs celebrate financial performance, hitting milestones, saving money, and similar goals. The key difference between incentives and recognition is recognition's connection between values and behavior.

Thus, incentives and recognition coexist as different management tools, addressing different critical goals of the organization. We might say that incentives are about hitting targets (left brain) and recognition is about applying values (right brain).

Recognition and Social Needs

The well-known hierarchy of human needs observed by psychologist Abraham Maslow provides a template for thinking about recognition's effectiveness and place in management's tool kit. Here is a simplified version of that model:

The First Pyramid—A Hierarchy of Human Needs Maslow's famous hierarchy of human needs begins with those related to physical survival (the most basic need) and climbs through the needs for safety, social contact, self-esteem, recognition, and status. The highest need (and psychological achievement) is called self-actualization.

Maslow's Hierarchy

Self Actualization

Esteem Needs
Self-esteem
Recognition status

Social Needs
Sense of Belonging
Love

Safety Needs
Security
Protection

Physiological Needs
Hunger
Thirst

Proponents of recognition in business point out that recognition satisfies the higher needs Maslow described. In fact, Maslow's pyramid can be seen as a metaphor for what a workplace can potentially provide, from the pay that assures food and shelter, to safety and social contact, to self-esteem. Self-actualization in the organization can be seen in those who love their work, who find their identity and satisfaction from their work, and who are "a perfect fit" with the organization.

The higher you climb Maslow's hierarchy, the more individualized the needs become. Physical needs are pretty much the same no matter who the person is; everyone needs food and water. Safety needs are more individual, but there are plenty of guidelines for creating a physically and psychologically safe workplace. Social needs require a workplace that functions socially in which company culture encourages socially productive interactions (peer-to-peer recognition is a socially productive interaction). Esteem needs and self-actualization are unique to everyone.

Recognition feeds the higher psychic needs of an individual to drive them to performance above and beyond just what's listed in their job description. When an individual feels his or her work is valued by a manager, that specific behaviors are rewarded, that the workplace is fulfilling the higher needs on the pyramid, engagement follows.

The Second Pyramid—A Hierarchy of Employee Needs. Jim Grenier, vice president of human resources, total rewards, and workforce solutions at software company Intuit, uses Maslow's hierarchy in an insightful model of employee needs. Viewing the relationship between Intuit and its employees, his hierarchy begins with basic needs for security and justice, then climbs to the needs for accomplishment, connection, and inspiration. Jim's illustration shows the connection between Maslow's individual needs and the relationship needs between employee and employer.

There is much management wisdom to be gained from this chart; for our purpose, let's note that Jim believes recognition of the right behaviors is a tool for promoting, encouraging, and confirming *all* of these needs.

It's no wonder that Intuit—where in an internal survey 93 percent of employees agree that the company's recognition program

Intuit's Model of Employee Needs

Inspirational Needs
"I see how our work makes my world a better place."
Identity and Meaning

Connection Needs
"I am a trusted, integral member of a community of people I like and respect."
Relationship and Belonging

Worth Needs
"I am challenged and capable of doing my work well and feel appreciated for my efforts."
Accomplishments and Esteem

Basic Needs
"I believe the company cares for my well being, providing for the basic, physical, economical and psychological workplace needs of employees."
Security and Justice

helps motivate sustained high performance —has been a "Best Place to Work" on the annual *Fortune* magazine list for years.

The Third Pyramid—A Hierarchy of Culture Management. Strategic recognition leverages the insights of these models to manage culture. Here's how it works:

1. Personal recognition satisfies the higher social needs of Maslow's hierarchy. *Recognition supports self-esteem and self-actualization* for individual employees. It encourages unity of values and unity of purpose.

Culture Management Hierarchy

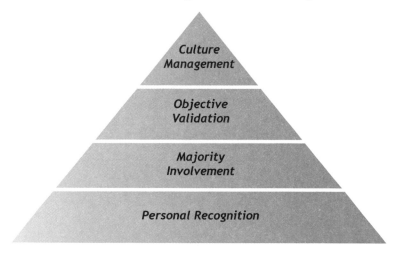

Culture
Management

Objective
Validation

Majority
Involvement

Personal Recognition

2. The next level up answers the organizational need to promote culture across locations, across time, and among many people. In this step, recognition is the key to *frictionless, positive reinforcement of desired behaviors across the entire organization.* Thousands of individual recognition moments promote, reward, and reinforce behaviors that advance organizational goals. Psychic income is abundant. Web technologies facilitate cross-communication. Participation by a large majority of employees—80 percent or more—guarantees that the culture will be self-sustaining because the behavior of that many *is* the de facto culture. (Conversely, if only 10 percent of employees participate in affirming culture, they reflect the behaviors and attitudes of a minority, not the true culture of the organization. We'll say more about this in Part II.)

3. Management science requires *objective validation* of outcomes. How do you prove that the culture, or one component of the culture,

is driving results? For this, the organization must have tools that allow impartial measurement and irrefutable correlation from value to behavior to outcome. Tracking recognition reveals strengths and weaknesses in the management of behaviors among employees.

4. As recognition is integrated with other management tools, it becomes a key to *culture management*. Culture management, through recognition and its related tools, is the top of our pyramid because we believe it is the ultimate tool for business management over the long term. Culture management is the *direction of behavioral choices based on values*. It is a clear communication, training, and reward system, and recognition is one of its tools. In fact, culture management is implicit in every organization. The relevant question is, will culture management be deliberate, measurable, and globally directed toward shared, transparent values? Or will it be haphazard, unmeasured, and individually directed toward vague or hidden agendas?

Think of it this way: Financial management achieves organizational goals through accepted and predictable accounting practices. Sales management achieves organizational goals through accepted and predictable contract terms. Culture management achieves organizational goals through accepted and predictable application of values in the workplace.

Practicing Values

David's best research and development group consisted of thirty software engineers in a virtual team spread across three countries and multiple time zones. This was the team he'd assigned to create a plan for SaaS that could migrate Carametric's services to a cloud computing foundation. The group posted daily reports of progress, ideas, and suggestions on a private company blog, and it was this blog that David read every morning. It was his informal way of monitoring the group's progress between biweekly reports.

A few days after the meeting with Tom and Laura, David reviewed the blog in his usual morning routine. A post on the blog caught his eye. It was from Mark, the team manager in the U.S. office, asking the common "buy or build" question to the group—should they buy a vendor's software package, or build the capability themselves? Replying to the post, a woman named Anka in the Poland development center proposed some interesting new ideas.

As David read through Anka's proposals, he felt an enthusiasm that had been familiar in his early days in technology. Anka had come up with a truly clever solution to the question, and it looked like she was already finished with a prototype. David scribbled "Innovation!" on a Post-It note.

David called Mark's cell phone. "Hey, get over here, buddy. I want to ask you about someone on the team in Warsaw."

✳

The mediocre manager, contemplating paychecks going out every few weeks, likes to think that his or her employees should be grateful to have a job. Perhaps they are, but that attitude has culture management backward. In a well-run company, the organization and the individual manager acting on its behalf harness the power of appreciation not by receiving it, but by giving it to the employees.

Let's look at the aspects of appreciation that make it essential to culture management:

Appreciation is motivating. People like being thanked. It feels good, affirming their worth and value. How do they get more thanks? By repeating the behavior that wins thanks.

Appreciation is humanizing. The ability to express appreciation is a key strength in a leader. Appreciation is an emotion that, in many cultures, lends power to someone else, in the expectation that they will receive it. Can you imagine having your thanks rejected? It makes the person saying "thank you" a little less exalted, a little more human.

Appreciation is specific. "Thank you" is reacting to a specific act, achievement, or attitude that's recognized in the transaction. It also lends credence to the importance and value of that act.

Appreciation is empowering. First, appreciation empowers by affirming the power of the individual to make a choice. (I don't have to earn your appreciation, but I choose to.) Second, because appreciation can be expressed by anyone in the hierarchy to anyone else in the hierarchy, it is a reward that potentially cuts across the class and culture lines of an organization.

Appreciation is powerful. Spiritual leaders emphasize the importance of gratitude on the path to wholeness. National leaders thank soldiers for their service; mayors offer the thanks of a grateful public to first responders to emergency situations. And notice how often the most enlightened business leaders attribute their success openly and often to their employees.

Appreciation establishes a psychological contract between employees. Complete that contract and you are assured more productive relations among workers. Break that contract and you are assured higher turnover, lower engagement, and a population of employees who deliver below their full potential.

Recognition as an Indicator of Company Health

It's ironic that many executives who accept the power of branding, which is an appeal to customer emotions, ignore the power of employee emotion. Perhaps the intangible benefits of work—like a sense of belonging, or a sense that one's job brings meaning beyond a paycheck—seem too easily dismissed as "just touchy-feely stuff." Human capital management, however, means applying management methodology to the emotional needs and power of humans.

We believe that recognition is particularly adaptable to the goals of management. Recognition done right improves retention of key employees, improves performance of all employees by guiding behavior, motivates the already-engaged employee to deeper engagement, and inspires both the recipient and others who witness the accolade.

Recognition also serves as an early warning system in a large organization, because it tracks optimum performance. When recognition is continuously tracked by management and executives, it reveals discrepancies and disconnects between managers and their staffs. For example, if recognition is not happening in a department, that's an early signal that something's going wrong. Is performance substandard? Is the manager unable to notice or reward the right behaviors and outcomes? Conversely, what if a department is performing magnificently but the manager is not recognizing or rewarding employees? That points to a potential risk of losing employees, or a manager who is out of touch.

Using company values as the reason for recognition also allows management to track understanding of these values by individual, team, division, region, or company-wide. For example, if the value of teamwork is measured across divisions, and one division is found to be well behind others in its acts of teamwork, then management can

MYTH BUSTER

✳ ✳ ✳

Are you only *retaining* employees or are you creating loyal employees? What's the difference?

"A survey by the Center for Work-Life Policy, an American consultancy, found that between June 2007 and December 2008 the proportion of employees who professed loyalty to their employers slumped from 95 percent to 39 percent; the number voicing trust in them fell from 79 percent to 22 percent. A more recent survey by DDI, another American consultancy, found that more than half of respondents described their job as 'stagnant,' meaning that they had nothing interesting to do and little hope of promotion. Half of these 'stagnators' planned to look for another job as soon as the economy improved. People are both clinging on to their current jobs, however much they dislike them, and dreaming of moving when the economy improves. This is taking a toll on both short-term productivity and long-term competitiveness: The people most likely to move when things look up are high-flyers who feel that their talents are being ignored." [xiii]

Employees agree to be retained in a tough economic environment or in other situations in which options may be limited. But if you're not fostering employee loyalty, as soon as more options become available, you will see your employee retention numbers plummet.

target intervention in the form of additional training, mentoring from other divisions and the like, all focused on increasing teamwork.

Recognition and Human Resources

For the organization, recognition is an accelerator of good human resource work. Specifically, recognition enables the appropriate direction, review, and strategic alignment of employees.

Recognition helps recruiting. Top candidates look at the company's culture in deciding whether to join an organization and are generally skeptical of management claims (they've heard it all before). They tend to seek information about the workplace through informal channels like social networking sites and their own professional networks. A company employing recognition will send a strong positive message through the employees that "This culture is authentic." In a company lacking recognition of accomplishment, human resources executives will see a higher employee turnover rate as talented workers reject the traditional, command-and-control management culture.

The human resources department is most often tasked with managing recognition programs, and here the old truism "If you can measure it, you can manage it" should guide its thinking.

Far too many human resources professionals are relegated to only those traditional practices that offer little competitive advantage. While important, the provision of benefits, compensation, workplace safety, and other factors are set by budgets and the competitive marketplace. As a result, HR is unjustly thrown into the cost center shooting gallery, its expertise and scope always under scrutiny. [xiv]

Human resources can offer competitive advantage by striving for excellence in the practices surrounding hiring, training, and retention, and recognition is a component of all three. Recognition

✳ Recognition and Talent Management

A perspective on the importance and power of recognition is reflected in a 2007 McKinsey & Company study[xv] that cited the top seven obstacles to good talent management:

1. Senior managers don't spend enough high-quality time on talent management.
2. Organization is "siloed" and does not encourage constructive collaboration and sharing of resources.
3. Line managers are not sufficiently committed to developing people's capabilities and careers.
4 Line managers are unwilling to differentiate their people as top performers, average performers, and underperformers.
5. CEOs and senior leaders are not sufficiently involved in shaping talent-management strategy.
6. Senior leaders do not align talent-management strategy with business strategy.
7. Line managers do not address underperformance effectively, even when chronic.

Recognition—done right—can overcome ALL of these obstacles.

burnishes the employer brand, thus attracting talent. Recognition is a form of ongoing training that highlights the right behaviors. Recognition aids retention and loyalty by paying out psychic income. Finally, human resources can become mired in the "irony of scarcity," a concept described by Steve Kerr, former chief learning officer at both General Electric and Goldman Sachs, in his excellent book *Reward Systems: Does Yours Measure Up?* For example, compensation and benefits are scarce resources, so HR professionals spend much of their work time deciding how to allocate them. Recognition, encouragement, and acknowledgement are flexible, limitless resources. There is certainly a financial value to a formal recognition program, but recognition does not have a single price tag. A "thank you" is free. A symbol of appreciation such as a gift card is far less costly than a raise or a better health plan. The irony is this: Because recognition isn't a scarce resource, it tends to be ignored. It's not screaming for attention, it's not threatening to get out of control, the investors aren't questioning it during the quarterly call, so the professional managers in HR and management aren't focused on it.

Related to this is the power of a quota. Put a quota onto a program, and managers pay attention. We believe that managers should meet a quota of employee recognition because that signals the importance of the program as well as the significance of finding recognizable behaviors. (We'll show you how to implement and manage this in Part II.)

Recognition is a management practice as flexible to the situation as the expertise of the manager applying it. It changes the game from the typical performance and compensation review conversation: "I've got a three percent budget for raises. You're getting 2.25 percent because these other people are performing very well. I'll see you next year." Depending on the merit structure of the company, managers may be required to operate in this way. But this structure sets up an inflexible discussion that does not account for the employee's motivation or engagement. Recognition allows many more flexible options for review and rewards the other 364 days of the year, depending on in-the-moment performance.

MYTH BUSTER

✳ ✳ ✳

The annual or biannual performance appraisal process is not the most effective means of conveying praise or constructive feedback to employees due to several limitations, not least of which are:

1. Because of their infrequency, appraisals are usually a source of anxiety for both the appraiser and the employee.
2. Standard appraisals primarily offer the viewpoint of one person with no real benchmark beyond the immediate team.
3. Appraisals give an imprecise picture of division performance.

Strategic recognition dramatically enhances the performance of employees by encouraging peers and managers to frequently and in a timely way acknowledge efforts and achievements that demonstrate the company values and contribute to company objectives. It is critical that the recognition come as soon as possible after the effort or achievement being rewarded.

These "recognition assessments" and kudos can then be used during the annual performance review as additional data on an employee's strengths (John has been recognized repeatedly for innovation) and even weaknesses (but John has been recognized only once for teamwork) and to identify potential areas of improvement. This presents a much more rounded view of an employee's contributions, some of which may not have been seen directly by the employee's manager. Moreover, since such a strategic recognition program is deployed company-wide, data can be gathered and used to benchmark an individual's performance and demonstration of values in their work against direct peers, team members, the division, and even the company as a whole.

Once a year, once a month, or once a week. Which do you think is going to have a greater impact on your daily behavior?

Peer Recognition

Recognition by peers is one sign that the company's culture has spread from the elites to the majority. Colleagues at any level of an organization bond with their peers, associate with them, share their successes and their obstacles. To institute peer recognition is to empower coworkers to honor each others' achievements, which is a powerful and cohesive force. When peers recognize each others' contributions, they build trust. Silo walls fall and information flows more freely. Peer recognition feeds psychic income needs and boosts morale while also relieving managers of the pressure to stay close to everyone (peers often know one anothers' contributions better than the boss).

Because this is about professional recognition, not personal popularity, peer recognition, like other kinds, requires the discipline of management practice underlying it. This is another way in which a formal recognition program contributes to better managed departments.

Who doesn't enjoy honoring a colleague? It's empowering and

builds general good feeling. When most employees participate, the company acquires a precious asset: a company-wide culture of appreciation.

A Culture of Appreciation

Recognition cultivates a culture of appreciation among employees, management, and executive leadership. That culture merges into the existing company culture; it becomes part of the thousand acts a day that define employment. Since we're recommending exactly this, let's briefly consider what a long-term culture of appreciation, as opposed to single recognition moments, might mean for a business.

A culture of appreciation creates an expectation of psychic income: If an employee is doing the right thing, it will be noticed, honored, and appreciated (as well as compensated in tangible ways). This expectation motivates each person to consider what values-based behaviors will earn that psychic income, just as an expectation of promotion causes some to take risks, perform beyond their job description, or seek new skills. The culture of appreciation can be the uniting force across the inevitable "silos" and departmental cultures, for while highly valuable behavior in one group might not be valued in another, the fact that both are recognized as appropriate to their situation is universal.

For example, risk-taking might be appropriate in design or marketing, where creativity holds the key to great work, but it might be shunned in accounting, where predictability and precision are key. Appreciation can honor risk-taking in marketing, and risk protection in accounting—the fact that appreciation is given is the relevant point.

A culture of appreciation also aids the creation of a robust social architecture in which communication flows freely, consistently, constantly. Employees are universally encouraged to do their best—they are not just complimented, but acknowledged when their behavior is aligned to company values and delivers strategic objectives. In the global organization, the real differences among cultures are celebrated by a culture of appreciation, overcoming the alienation and miscommunication that is the hobgoblin of a multinational enterprise. It encourages trust and that almost mythical bonding that soldiers call "unit cohesion."

*

In summary, recognition uses a company's social architecture and a careful payout of psychic income to increase employee engagement and performance. That recognition should go beyond the moment between a manager and employee, beyond the single relationship of worker and boss. A culture of appreciation is self-sustaining and helps everyone live the values of the organization. Many fine companies find this culture of appreciation such an improvement to their organization that they pause after implementing enterprise recognition, satisfied.

There is a much greater potential implementation of recognition, however, that brings all the acts of appreciation and motivation into the most rigorous practices of management, on a par with financial, legal, and operations management. That is the final evolution of the practice, strategic recognition.

* * *

✳ ✳ ✳

Chapter 3
STRATEGIC RECOGNITION

Just before leaving for the night, chief technology officer David arranged for an award of $100 to be sent to his colleague, Anka, in Poland. He wrote:

"Anka, I want to thank you personally for taking over the validation module for the SaaS project. I have read the report from Quality Assurance and they tell me, as we say here in the U.S., it is rock-solid. (How do you say that in Polish?) We might have used an off-the-shelf module, but if what I'm reading is right, your team built something that we might use across the company. That's innovation! Please accept this award and my congratulations. I hope it reminds you that people like you make the difference."

David hit the "send" button and shut down his email. He looked at the row of 24-hour clocks on his wall. It was 3:00 A.M. in Poland; even Anka wouldn't be awake. The recognition award would be the first thing she saw in the morning.

David slid his laptop into his briefcase, turned out the office light, and walked toward the elevators. A few engineers were still working and he stopped to thank them, to chat about their work. He thought about how easy it seemed now to keep the team running at top performance. Easy and, he had to admit, even fun.

This chapter is for executives. It shows how the management practice we call *strategic recognition* can drive individual and team performance to achieve your strategic objectives. Here also is our vision of recognition as a full partner in the most sophisticated and innovative management science, as distinct from old-fashioned recognition practice as Six Sigma is distinct from "good enough" quality control.

Why "Strategic" Recognition?

Strategic recognition is all about delivering on your business goals. It helps employees understand the behavioral norms you have identified to achieve the desired business outcome—and it does this for every employee in your company, regardless of where they live and work.

To understand the power of this concept, it's helpful to examine those two words separately.

Strategic suggests that if you do recognition right, you will care about this program beyond its boost to morale. A truly strategic program gives the CEO new insights into the behavioral norms and the culture driving the company. Appreciation will have a positive and measurable effect on productivity.[xvi] This means your system enables you to *monitor, measure, and manage* investments you're making in recognition, compared to discretionary spending that doesn't drive strategic outcomes.

Recognition in the strategic context means constant reinforcement of strategic values. For example, if quality is a corporate value, you must be able to give "quality" a rich meaning applicable to the way your employees behave every day. Recognition isn't just about delivering awards. The recognition system should help achieve critical strategic goals but shouldn't interfere with the very real human emotions that go along with recognizing and being recognized for positive behaviors.

In fact, we believe that old-fashioned recognition's emphasis on giving away a coffee mug with the company logo impinges on that positive moment by adding an object to an interaction. If the award is meaningful but the system is invisible, then the recognition moment becomes two people showing mutual respect, common values, and support. This is a much more human and intimate moment that may seem a little scary to some. What if, instead of focusing on the glass paperweight with the employee's name engraved on it, the moment was about real human appreciation? What if the focus was on the special wording in the recognition moment, and the material reward was chosen later by the employee? Managers would have to drop the organizational mask

MYTH BUSTER

✳ ✳ ✳

Old-fashioned recognition's emphasis on ceremony implies that managers have to perform a kind of "appreciation ceremony," a moment that will be energetic and fun, like a popular game show.

We disagree. People show appreciation in different ways depending on their temperament, cultural upbringing, and perception of what's appropriate in the workplace. What matters in the recognition moment is not showmanship but sincerity. Let the managers express appreciation in their own way; if it's personal, tied to company values, and genuine, it will be effective. This is also a far more respectful approach for employees who may view public attention—even for recognition—as punishment and not reward.

and show themselves as sincerely appreciative individuals! What would that do to the company's culture?

The Five Tenets of Strategic Recognition

Five operating principles, or tenets, drive successful global strategic recognition throughout the organization. These tenets prevent the managerial and psychological "disconnects" that plague less effective programs and guide the ten-step framework for implementation discussed in Part II of this book.

TENET 1:
A SINGLE, CLEAR GLOBAL STRATEGY

Global means universal, whether you work for a small organization or a multinational giant. A clear global strategy requires a clear outcome. Some examples of the outcomes that might be sought are increased customer satisfaction, increased employee engagement scores, more repeat business, higher net promoter scores (the metric created by loyalty expert Fred Reichheld), and a measurable increase in product quality or reduction in costs.

✳ Research Insight

"A company's success depends on its people. But their collective power stems, in part, from an organization's ability to point them in the same direction and, importantly, in a direction that is aligned to the organization's business strategy. When an organization's leadership, workforce, and culture are aligned with its strategic priorities, people can be a major source of sustainable competitive advantage."

—TOWERS PERRIN [xvii]

Many clients have come to us with multiple, scattered, and even conflicting recognition programs, which leads to divisions within the company, confusion, and wasted money. A global strategy creates a single recognition brand and vocabulary. It creates clear visibility into budgets and can be audited. Executives in different divisions, locations, and markets can view uniform metrics that provide insight into program adoption, operation, and results.

You have to treat all employees equally. We have an expression across global companies, *parity of esteem*, which means that whether employees are in Ireland, Poland, Japan, or Argentina, they are all treated with the same regard.

Equal treatment does not mean identical treatment, however. A clear global strategy includes recognizing the differences in languages and local cultures and assigning award values to align with the local standard of living. Rewards must be personally meaningful, culturally relevant, and equitable in the number of options and value of rewards from country to country.

If a company already has ad hoc recognition programs in place, the transition to a single strategy will be challenging. Some programs will have to be cancelled, and some people will be very attached to their ideas of recognition. To ensure consistency as well as manage the program, you have to integrate a single technology platform. You have to establish a return on investment in new technology across divisions (and factor in the savings of canceling the ad hoc programs and consolidating everyone on one platform).

TENET 2:
EXECUTIVE SPONSORSHIP WITH DEFINED GOALS

Support from senior management is critical to success in any initiative, and this is especially needed in managing corporate culture.

In market-leading companies, strategic initiatives are managed using process, metrics, incentives, and accountability, and senior executives monitor these. Success requires a rigorous methodology (such as Six Sigma's DMAIC [xviii]) and meaningful measurement. Managers must be held accountable for goals, including percentage of employees awarded, employee satisfaction and engagement scores, the match of award distribution to the bell curve, and the frequency of awards. Sponsorship and communication of the recognition program cannot be left to human resources staff; it must come from a group of unified top leaders—all of them.

As we'll see in Part II, the more exactly you can plan such details as award levels, the more precise your measurements of award activity and money spent will be. With a defined goal of 80 to 90 percent participation in the recognition program, your data will be actionable. With defined participation goals for division heads and managers, the cultural management of recognition can take hold.

✳ **Executive Insight**

"Our culture is based on the fact that people have an innate need for well-deserved recognition. Using recognition is the best way to build a high-energy, fun culture and reinforce the behaviors that drive results.

"[Recognition] needs to be deserved, and it needs to come from the heart. So I think what leaders have to do is recognize the people who are getting it done. For the people who are getting it done, [recognition] can't be done too much. Why be selfish on the thing that matters most to people?

"People leave companies for two reasons. One, they don't feel appreciated. And, two, they don't get along with their boss. We try to recognize the people who are really getting it done any time they happen to get it done, and we try to develop coaches instead of bosses. To me, with recognition, if you've got to err on the side of anything, recognize more than you should."

—DAVID C. NOVACK,
Chairman, Chief Executive, and President of Yum Brands [xix]

MYTH BUSTER
✳ ✳ ✳

Recent corporate history is strewn with examples of executive and employee malfeasance driven, largely, by the desire to increase personal gain through a company-sponsored bonus or reward program. As a result, many have denounced reward programs entirely as a means for enabling and encouraging corruption. This is where the "strategic" component of recognition becomes critical or you could end up with just this kind of negative behavior. You must positively reinforce only those employee actions that reflect the company values while achieving the firm's strategic objectives. This approach ensures that employees who, for example, increase productivity but do so by harming the environment will not be rewarded for their efforts. Values-based recognition is the key to ensuring that employees display the right behaviors in achieving company goals.

TENET 3:

ALIGNED WITH COMPANY VALUES AND STRATEGIC OBJECTIVES

As we discussed, when individual-recognition moments are consciously linked to company values and goals, employees understand how their actions directly affect the culture. They see how their behavior fits into the big picture. They gain both a sense of efficacy and a sense of accountability within the big picture.

For all this to work, you need to track the program in a far more rigorous and disciplined way than the usual recognition effort attempts. If you can count the number of times in a company that somebody thanks somebody else for going the extra mile on a value like quality, that can give you an indication as to the amount of discretionary effort that's being expended in and around quality. Management science suggests you then add that all up and you accumulate this information. You put it together graphically in a histogram and you compare a quarter against a quarter, a country against a country, a division against a division. This can give you enormous insight into how your values are turning into behaviors and are displayed in discretionary effort across the company.

Monthly dashboards illustrate for managers the traction of each value, whether by region, division, or department. Targeted management intervention in places where values are ignored or misunderstood then becomes possible. The dashboards represent people's behavior, which as we've said is the reality check of a company's culture.

We'll illustrate these histograms in Part II. For now, remember that the central point of recognition is to identify and encourage alignment with company values and strategic objectives.

Encourage Alignment
with Values and Objectives

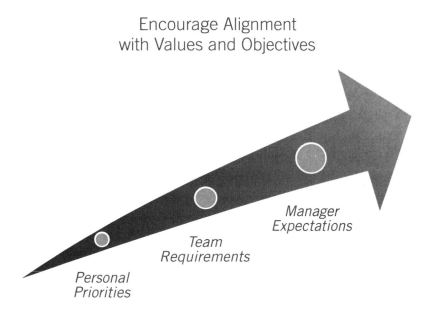

*Manager
Expectations*

*Team
Requirements*

*Personal
Priorities*

TENET 4:
OPPORTUNITY FOR ALL TO PARTICIPATE

When only a few elite members of the organization receive infrequent, high-value awards, it is impossible to affect the broader corporate culture. Giving many lower-value rewards to employees across the company, by contrast, results in a stronger impact on the company. Every recognition moment doubles as a marketing and communication moment, reinforcing company values in a positive employment experience. As more employees participate, the company gains greater voluntary alignment with shared values. As more participate, the data described in Tenet 3 above become richer, more detailed, and more accurately reflective of the entire company's attachment to particular values. As a side benefit, the presence of a broad-based recognition practice breaks down psychological barriers of class and rank, decreasing the chance of employee alienation from management.

MYTH BUSTER

✳ ✳ ✳

What would boost your engagement at work? The answer is surprisingly different if you're sitting in the employee's seat or the manager's. In a study developed by OfficeTeam and the International Association of Administrative Professionals (IAAP), the disconnect couldn't be more pronounced. Administrative professionals cited an "in-person thank you" as their most desired and valued form of recognition, tied with "boss shares achievement with senior management." Managers, however, ranked these options last, after promotion, cash, and paid time off. [xx]

A simple thank you, sincerely and meaningfully given, carries much more weight with staff than many monetary rewards. Managers must not fool themselves, however, that they are communicating appreciation to their employees effectively. In another recent study conducted by the Institute of Leadership & Management, 56 percent of managers said that they appreciate their employees, but only 23 percent of staff agreed. [xxi]

TENET 5:

THE POWER OF INDIVIDUAL CHOICE

The relevance of an award to an individual is more important than its material value, especially in a global program. When developing the roster of awards available, managers must consider the demographics of a worldwide workforce that might span four generations, all with different expectations and driving forces. Locally based choice ensures the award will always be culturally appropriate and to the recipient's taste while avoiding the varying cultural norms that simply cannot be known by every manager everywhere in the world. Allowing people to choose what is meaningful and personal to them increases the significance of the award. (Movie tickets aren't motivating for someone who doesn't like movies. A cupcake won't motivate someone on a diet. A designated parking space means nothing to someone who rides the bus.) Noncash rewards in the form of gift cards to local high-value venues take rewards beyond compensation to a socially acceptable trophy status everywhere in the world.

MYTH BUSTER

* * *

What have you received for rewards in the past? We've heard horror stories of people receiving clocks in an Asian culture where that symbolizes death. We've heard about the reward recipient in a European country actually having to go to his local customs office, fill out numerous forms, and then pay customs duties out of his own pocket—just to get his "reward." We've heard about grandfather clocks used as rewards and shipped from the United States to recipients around the world, including Australia, costing the company more in shipping than the "reward" was worth.

Individual choice of rewards also aligns well with the expectations of younger talent, the future leaders of your organization. They are savvy about media and brands, not shy about expressing preferences, and conditioned by technology and temperament to express their preferences.

A System of Total Rewards

"Can't we just give cash bonuses?"

Giving cash rewards for good performance is a severely limiting form of recognition and motivation. To make the point, we'll call on the total rewards model as described by management consultancy WorldAtWork, whose recognition practice leader Alison Avalos told us the following:

"Compensation is the *must* in the package, but in the end rewards are not only about pay. Recognition can be used to meet specific needs to differentiate one company from the next. Recognition is customizable, informal, and easy to shift based on what you're trying to accomplish. Most high-performance companies have figured this out. What works today might not work in five years; typically a workforce is going to evolve. Today, flexibility is at the center of the map."

A total rewards system includes compensation, benefits, work-life factors, performance and recognition, and development and career opportunities. It includes both tangible income and psychic income. The goal of total rewards is to achieve the highest return on investment with the optimal mix of rewards. In practice, managers calibrate and apply this menu of rewards to attract, motivate, engage, and retain employees individually. That leads directly to improved performance and business results.

Recognition caters to the psychic investment you make in yourself and others make in you. That individual focus has to carry through in an award that crystallizes the message. Recognition is often forgotten, however, by staff at all levels who typically only see pay, benefits, and sometimes equity in the company reflected in their compensation statements. This is a problem of visibility not value.

In *Reward Systems: Does Yours Measure Up?*, human resources pioneer Steve Kerr observes that compensation, benefits, and incentives have an easy-to-measure cash value. He goes on to describe what he calls *prestige awards* when, for example, an employee becomes a member of the prestigious President's Club or gets the window office. Kerr also describes a category of awards he calls *content rewards*, which include feedback, conversations between managers, and recognition, either one-to-one or publicly shown. These content rewards are a continuous performance lever to reinforce the culture on an everyday level. (Other content rewards can include giving an employee a new role, exposing him to training, or nominating her to an important committee.)

A strategic recognition program is paid for in cash, but its accretive value, memorable human connections, and built-in flexibility make it more adaptable to the goal of supporting values. In a well-designed program, recognition is more abundant than cash because, typically, a recognition award has an average value of $100, whereas a base salary might be $60,000 or a bonus might be $5,000. So that $100 is obviously much less scarce than a $5,000 bonus.

This begs the question: should we think of recognition as fundamentally a cost outlay, like the cost of employee health insurance, or an investment in the quality of management? Companies that have made a success of recognition see it as part of the "total rewards" view of compensation because it helps management succeed and optimizes performance, delivering an impressive return on investment.[xxii]

Cash vs. Noncash Rewards

Ask any employee if they want cash or some other reward, and they will nearly always say cash, believing it gives ultimate flexibility. That might not benefit the organization, however. If your investment in recognition is nothing but cash prizes, you're making an investment that gives you the worst possible return for your investment. To understand why, let's look at where that cash goes.

Let's say you receive $100 cash in a spot recognition program. How would you use that award in your personal life? It would probably be distributed in your paycheck as just a line item on your paystub and then deposited into your household checking account. Income taxes are taken out of the award amount, so the award gets reduced to about $75.

An employee making $50,000 a year receives a biweekly before-tax paycheck of $1,923.07 and an after-tax (at a 25 percent tax rate) deposit of about $1,442.31. Does a one-time change in an employee's pay from $1,442.31 to $1,517.31, two weeks after the recognition moment, seem memorable? We doubt anyone would notice.

Studies with our clients show that for those using cash awards in their recognition program, many (and sometimes most) of the employees who receive cash awards have no recollection of how they used those awards.

Why is this? Because cash is slippery. Welcome as it might have been, the psychic payoff was fleeting, and the award was undistinguishable from other compensation. In Maslow's hierarchy, many cash awards end up going to pay for life's necessities—the lowest part of the pyramid.

Now let's look at what happens with a noncash reward: You receive a $100 gift certificate in a spot recognition program that allows you to choose how and when you want to redeem the certificate. How would you use that in your personal life? First of all, you would likely have to take some time to decide how to use the certificate, already making this reward more memorable than the cash award described in the previous example. In this instance, the award remains separate from your paycheck, and gives more discretionary purchasing power. You can choose the award's emotional content. You might make an impulse purchase without guilt. You might buy a special gift for a friend. You might give the card to a charity important to you. Every one of these decision points prolongs the psychic payoff. And at every point in the decision process you remember, "I got this because I demonstrated innovation in meeting that deadline."

✳ Research Insight

"The traditional forms of motivation are compensation and benefits. The problem with these tangible rewards is that they are short-term motivators. The more people get, the more they develop an entitlement mindset. Adding more and more tangible rewards does not necessarily increase motivation or engagement. However, taking away tangible benefits or entitlements really de-motivates or disengages people.

"On the other hand, intangible rewards, such as a 'thank you,' 'good job,' or effective coaching, let people know their managers care about them and value their contributions. The more intangible forms of motivation the better—they raise engagement levels by helping people feel connected.

"The additional advantage of using intangible rewards is that while offering them greatly increases levels of engagement and motivation, withholding them tends not to have a significant long-term de-motivating impact. Additionally,

intangible forms of motivation are not costly to provide. So for a small investment of time in showing appreciation, the resulting improvement in engagement and connectivity can be huge. The key is in giving credible, sincere, and respectful appreciation."

— KEVIN J. SENSENIG,

"Human Potential Untangled" [xxiii]

The Need for Measurement

Business measures activity for understanding and control: "If you can measure it, you can manage it." Measurement is a reality check. If your recognition program is to achieve that level of strategic cultural management, you have to monitor its activity day-to-day.

To manage the success of a recognition program, you need to measure the act of recognition itself. This can be done by giving your managers targets and having part of their management by objectives (MBO) bonuses depend on hitting those targets. For example, "You have to give ten awards to your team this quarter. It's in your objectives. And I'm going to measure that."

In practice, most acts by employees that are to be recognized are not things that are formally measured against preset goals (that would be acting on incentive). These acts are spontaneous and depend on the employee's good judgment. They are inspired from within. They are a surprise.

Here's an example: Globoforce office manager Kim organized a daylong off-site meeting. She did a fabulous job. Nobody sat down with Kim beforehand and said, "If you organize the off-site by 5:00 on Friday, and we all get our meals on time, and you go the extra mile to make it a success, then you're going to get an award." That would be an incentive. Instead, Kim made a series of decisions on her own that ensured the meeting ran smoothly and was successful. That's the kind of behavior that should be recognized (and it was!).

With incentives, managers are encouraged to reward what's measured; with recognition, they are encouraged to reward what matters subjectively and spontaneously.

✳ Executive Insight

"It is better to use imprecise measures of what is wanted than precise measures of what is not."

—RUSSELL ACKOFF,

Seminal Author and Educator on Management

Systems, Known as the "Father of Operations Research"

Is it contrary to the spirit of recognition to place a certain number of recognition moments in an MBO plan? Doesn't that make acting on recognition an incentive? Leaving aside the fact that you can "recognize the recognizers," we believe that making recognition a requirement elevates its status to that of a strategic practice. Recognition should be in managers' MBO targets.

A common reply to this concept of having a quota of awards in your MBO plan is, "What if our managers just don't see that much good behavior?" And our response is, "Maybe they should open their eyes. It's almost never because good behavior wasn't there. It just wasn't being noticed and recognized."

In *Reward Systems,* Steve Kerr observes that many "rules" of rewards are actually the cause of dysfunctional reward systems, because they engage in "the folly of rewarding A, while hoping for B." [xxiv] In one common example, he cites the company that wants innovative products and new ideas, but rewards employees who don't make mistakes. The company's reward system is encouraging risk-averse behavior, not innovation (which is inherently risky).

Participation and Penetration

In recognition programs, broad participation is essential for success. A recognition program will not foster the desired culture, behaviors, and results if only 10 percent of employees receive an award once a year because that's simply not enough penetration to raise awareness of the behaviors the company is trying to promote. Again, since the people are the culture, you're not going to change the culture by affecting one in 10 employees. In fact, if the recognition is only hitting 10 to 20 percent of the workforce, management has lost touch with the company. At the very least, the company is telling itself that it's not performing very well!

Our best practices, which have been substantiated by research conducted by the Stanford Graduate School of Business, have found that if the recognition program is promoted so that five percent of the workforce receives a recognition award each week, a critical mass will be achieved and the program will both maintain and promote itself. Top-performing companies ensure that 80 percent of the global workforce will be touched by the program each year, including peer-to-peer recognition—one of the most powerful methods for driving this level of penetration.

✳ Research Insight

"The key to driving productivity gains is increasing engagement among core contributors, who represent 60 percent of the typical workforce. Highly engaged employees are already

*working at or near their peak but are often limited by their
less engaged coworkers. Focusing on engaging core contributors can improve both groups' productivity."*
—WATSON WYATT WORLDWIDE,
2008/2009 WorkUSA Report

High participation levels get employees involved in promoting the
cultural change among each other. Think of how eBay users have
the star ratings for buyers and sellers to regulate each other for trust-
worthiness and customer service. The CEO of eBay doesn't decide
whether a vendor is supporting the values of honesty, service, and
transparency—the users do. In the same way, wide participation
and peer-to-peer interactions through the recognition program support the values promoted by recognition. Exceptional employees are
recognized by the group, and the group looks to them for informal
guidance. Executive management needs only to structure the recognition program so that it reflects critical values (or a big global
initiative), and the recognition program will provide incentive for
behavior that supports it.

Does Everyone Get an Award?

A common comment in reaction to recognition programs is, "Wait a
minute. Not everyone is equally deserving. Not every contestant gets
a trophy. If everyone gets an award, don't awards lose meaning?"

That would be true if global strategic recognition were a zero-sum
game, but this confuses the meaning of the recognition experience.

The goal of strategic recognition is to reinforce certain values
and behavior, not to make everyone feel good (the fact that it *does*
make people feel good is a benefit, but not the goal).

According to Towers Perrin's 2007–2008 Global Workforce study, "Companies with high employee engagement have a 19 percent increase in operating income and almost a 28 percent growth in earnings per share." For a company like Procter & Gamble, that means tens of billions of dollars in additional shareholder value. Engagement won't improve by 15 percent (a percentage that correlates to a two percent improvement in operating margin, per Towers Perrin) if only 10 percent of the workforce is getting continuous feedback on its performance. Achieving a 90 percent participation rate in a recognition program will cause an increase in engagement in a significant percentage of the workforce. Delivering a 15 percent improvement suddenly looks possible.

In real life, when our clients get to 80 to 90 percent penetration with their recognition programs, a bell curve of award winners appears. The lowest-performing 10 percent of employees will get zero awards, as is appropriate. The middle 80 percent might get two or three awards a year. The top 10 percent will receive perhaps six awards a year.

The people who win ten awards a year are a meritocracy, whatever their position in the company. They are receiving annual awards with an aggregate value of $1,000-$2,000. This is precisely the goal desired in a meritocracy. Top performers will be differentiated whatever their salary bands, whatever the budget for bonuses this year, because they lead your culture and the positive business results the culture is designed to deliver.

If only 10 percent of your employees feel like winners, 90 percent feel like losers (a year is a long time to wait for a "thank you" or even simple feedback on performance). Under such practice, there is a very small winners' circle, to borrow a phrase from Intuit's Jim Grenier. On the other hand, if you've got 90 percent of employees who feel like winners and only 10 percent who feel like losers, that is a much better mix. This draws a much larger and more relevant winners' circle based on merit.

If penetration is high enough, the recognition program becomes self-marketing because the vast majority of employees are winning awards and giving awards. This is the ultimate goal: Employees know about the program, interact with it, redeem awards, and are reminded of the values being promoted by the company. The positive emotional impact from winning recognition ensures that employees are more likely to participate. High participation is inherently efficient. Otherwise, continuous program marketing is necessary, reminding employees it exists, getting them to partici-pate—a death spiral that overtakes too many individual-recogni-tion programs. (Look at it this way: If your benefits were structured so that only 10 percent of employees had them, would the other 90 percent think of benefits as part of their total rewards?)

Work/Life Benefits

Any mature discussion of talent management in the last twenty years has considered the impact of work on the rest of life (e.g., work/life balance). Recognition's power to inspire self-esteem and align personal behavior with corporate values has implications for work/life balance that the alert manager will harness and exploit. At its most basic, the positive feeling that recognition fosters comes home as surely as negative job stress does. While it might be naive to say a good job guarantees a good life, a miserable job probably does more harm at home than good. In this way, recognition has a positive secondary effect on the most significant relationships in the employee's world—the relationships with family, friends, and community.

MYTH BUSTER

✳ ✳ ✳

Those employees known as millennials (also called Gen Y) are changing the work/life balance discussion forever. Millennials are not as concerned about balancing their work life and their personal life as they often do not clearly differentiate between the two. Work and personal time are so blended for them that relationships at work and the ability to work anytime, anywhere are important aspects of their day. For this and other reasons, millennials especially are seeking purpose and meaning in their work. A powerfully positive way to give them this purpose and meaning is by incorporating your company values and objectives into a strategic recognition program. This gives millennials a sense of purpose and accomplishment within the bigger picture while also emphasizing the importance of living the company values in their daily work.

Recognition takes management into a much deeper realm of human psychology than good feeling; its ability to keep an employee moving forward, growing, improving—and conscious of that growth—nurtures a deep desire for significance in what might otherwise be an impersonal corporate monolith. Think about this for a moment: If the organization encourages and enables a sense of significance—of all this work amounting to something more than a paycheck—then ties of real attachment grow. Loyalty grows. Other organizations that lack that intimacy lose their appeal to the employee.

What if you, the executive, could institute one practice that would carry your people through the good times and the hard times, the times when they feel accomplished and the times when they confront failure? And what if this practice was saturated with learning and genuine human contact? Among all the practices you must sustain, where would that fall in importance?

Recognition per se is not meaning or purpose, any more than plans are finished products; but recognition, like planning, is an enabling technology that helps employees create and experience meaning and purpose.

Recognition and the Marketplace

The third player in all employee-company relationships is the marketplace. Curiously, this is often neglected by managers when considering the impact of a new practice. Customers, vendors, business partners, shareholders, the media, stakeholders in corporate citizenship, and global geographic and online communities that interact with the organization all make up the marketplace that experiences the secondary benefits of strategic recognition.

Does that seem grandiose? We think not. Let's look at some outcomes of the changes recognition can enable:

* Participation in corporate citizenship, charity work, and community events
* Behaviors and decisions that maximize shareholder value while enacting company values (*double meaning intentional*)
* Cross-cultural information sharing, whether across divisions or around the world
* And most importantly, commitment to the customer on behalf of the organization, not just as an individual but also as a representative, indeed—the embodiment—of corporate culture. Jennifer

Reimert, senior director of global compensation at Symantec, follows this path when she says, "Employee loyalty drives customer loyalty, which drives revenue, making recognition a business proposition."

Strategic recognition is a positive player in the talent marketplace. It makes an organization more appealing to candidates and top talent because they look for an authentic and effective culture. The transparency of today's recruiting means that a company that appreciates its people and maintains high morale and high respect for achievement will have its pick of talent.

Strategic recognition is regarded positively by the press and opinion leaders, who look both at bottom-line results and management methods. Many companies tried in the 1980s to achieve a Six Sigma standard; GE did it and received worldwide praise and interest. Companies on *Fortune's* 100 Best Companies to Work list get lots of praise and interest for being on the list (and their return on investment beats that of the competition).

Strategic recognition, in sum, is a multidimensional practice with positive effects inside and outside the organization. It touches every other company initiative because it is focused on the all-important interplay of group culture and individual needs.

As a management practice, strategic recognition requires a framework for planning, launch, and continuous improvement. That's what you'll find in Part II.

Carametric, Eight Months Later

"R&D got its game back," said David. It was eight months since the meeting when Tom and Laura had challenged the R&D group's ability to innovate.

MYTH BUSTER

✳ ✳ ✳

Without your people you have nothing. They are truly your greatest competitive advantage. Yet survey after survey, article after article, show that the majority of employees are planning to jump ship as soon as they safely can—largely because of the way they have been treated by their employers during the recent recession. Deloitte Consulting reports that 65 percent of senior executives polled in May 2009 are highly or very highly concerned that high-potential talent and leadership plan to leave when the economy improves. Twenty-six percent of these executives report that they are seeing an increase in turnover of their best workers during the March–May time period. [xxv] Employees often understand why company leadership has to reduce headcount, cut costs, freeze pay, and other actions during an economic downturn. It's the lack of respect and recognition for what the remaining employees are able to do that is behind this mass desire to "find someplace where I'm appreciated." What are you doing to show you appreciate your employees today and value their contribution over the long term?

"You're right," said Laura. "I can see it in the values distribution graph."

"My employees are approaching 80 percent participation in the recognition program," said David. "The managers are hitting their award quotas. And R&D is buzzing like a hive."

Laura added, "The values distribution graph tells us the innova-

tion initiative is working, but I want to say a word about the quality of the work. The folks in customer care say that customer satisfaction on the early SaaS products is approaching 90 percent. Calls are the lowest they've ever been for a launch. Customer surveys show that 65 percent of people agree with the statement, 'Carametric's products are innovative.'"

Tom nodded with satisfaction. "This morning we got a call from the Wall Street Journal's technology reporter asking about the transition to SaaS. Dave, why don't you join me when we return that call? Oh, and Workforce magazine requested an interview with Laura about Carametric's drive for innovation. So even the press is noticing."

Tom concluded, "You know, I'm beginning to believe this culture thing." He smiled. "Good job, people."

✳ ✳ ✳

***

PART II
Implementing Strategic Recognition

* * *

Chapter 4

BUILD A STRATEGIC RECOGNITION PRACTICE

If an investor asked, "What is your compensation strategy?" you'd drop on the desk a three-ring binder filled with everything having to do with compensation: base pay, salary bands, grades, bonuses, all of it. If the investor then asked, "Explain your benefits plan," you'd have another three-ring binder stuffed with information about 401(k)s, health and education benefits, savings plans, profit-sharing, and stock options for employees. Next, the investor asks, "What's your strategic plan?" You roll out sixty-slide decks filled with product plans, customer profiles, marketing plans, and financial performance data. This planning is good.

When we ask executives, "Tell me about your recognition strategy," they have no binder, no deck of slides, no metrics, no execution plan. Compare this to all the practices named above. Recognition, for all its importance, is an outlier, not subject to the same rigorous planning and monitoring as other management practices. Why is this?

Done strategically, recognition can be planned and executed in a company like any other management practice, and therein lies the opportunity for competitive advantage.

Most of your competitors punt all responsibility for recognition to "Bob and Mary," mid-level workers in the HR department. Bob and Mary are not required to operate recognition as a strategic practice, yet recognition is one of the few practices left that offers competitive advantage. Others, like compensation, financial discipline, and quality control are so well established that your competition manages these in essentially the same way as you. If you elevate recognition to the level of other strategic practices, you create a fresh competitive advantage.

Part II of this book is a step-by-step guide to establishing a strategic recognition program and creating that competitive advantage.

Strategic Recognition Framework

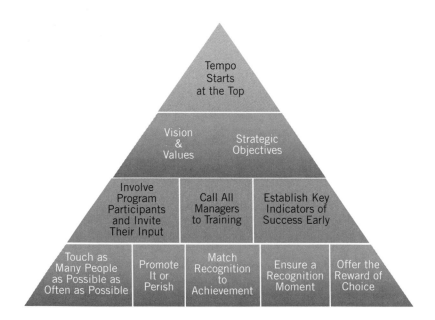

The Framework of Strategic Recognition

Earlier we discussed the five tenets of strategic recognition. Expanding on these, we have found that ten tactics constitute a full execution plan—a framework to make strategic recognition operate in your company. This approach of determining the strategic, then implementing the tactical has worked well for companies large and small around the world, and all actions can be adapted to your organization's unique culture. Follow them as outlined in this section, and you'll establish a powerful, proven strategic recognition practice.

TACTIC 1:

ESTABLISH PROGRAM GOALS AND OBJECTIVES

Why are you doing this? Without real, detailed goals, the practice of engaging and motivating employees can become "recognition for recognition's sake," another old-school program lacking accountability and relevance.

Begin with your organization's vision and values. What are the hallmarks of your culture? What do you want to accomplish? How does it relate to your most important business drivers? What must be supported—profits, customer service, operational excellence, innovation, and/or maximum product quality?

Often the goals of a recognition program begin with the question, "What do you want to change?" For example, do you want to increase employee engagement? If so, you need to increase both the commitment of employees to their work *and* their alignment with company goals. Commitment without alignment means wasted effort (and frustration). Alignment without commitment means wasted potential (and employee turnover). Recognition singles out great performance (commitment) that focuses on strategic goals (alignment).

A survey of companies by WorldatWork found these to be the most frequently cited goals for recognition programs.[xxvi]

RECOGNITION PROGRAM GOALS

✳ ✳ ✳

Create a positive work environment	81%
Motivate high performance	75%
Reinforce desired behavior	71%
Create a culture of recognition	70%
Increase morale	65%
Support organization mission / values	62%
Increase retention / decrease turnover	42%
Encourage loyalty	38%
Support a culture change	23%

Whatever your goals, they must be written down. As in any business planning process, supporting your strategic goals will be specific program goals and objectives. Here's an example from Intuit:

INTUIT'S RECOGNITION AMBITION

✳ ✳ ✳

* Drive higher employee engagement at Intuit.
* Improve employee satisfaction survey results.
* Create a culture of recognition throughout Intuit.
* Continue to make Intuit a "Best Place to Work."

Each of these goals can be affirmed over time—does employee engagement improve? (You can measure engagement objectively by such standards as the Gallup Q12 or Towers Watson models.) What do employee satisfaction surveys tell over time? Are line managers able over time to observe and describe changes in both commitment and alignment?

Ambitions may be stated as strategic goals for the recognition program, as in this example from Symantec:

STRATEGIC GOALS FOR RECOGNITION AT SYMANTEC
* * *

All recognition programs globally will be migrated to *one platform* with *one common brand* and *one executive dashboard*
* One global strategic recognition solution
* Drive employee loyalty
* Reward behaviors that support company values
* *Local impact* and *relevance* for all employees GLOBALLY

These are some of the most popular goals; yours must be specific to your organization's ambitions, market position, and challenges. Take time to define these goals clearly, for without them, recognition will not be taken seriously as a strategic initiative. (Without clear goals, moreover, you'll be tempted to judge your program's effectiveness by anecdote or gut feel. This almost guarantees program mediocrity.)

State goals as direct outcomes, even if factors beyond recognition come into play. For example, human resources professionals know that recognition can contribute to employee retention, even though retention bonuses, company culture, and personal ambitions also affect retention. In such cases, shared goals can be considered jointly against joint outcomes by tracking all relevant efforts.

Here's an example of the relevance of goals: We often say they don't teach recognition in management school, so most managers are ill-equipped to make use of it and must be trained. More often than not, managers avoid a new behavior like recognition. That's why they need goals. Measurement against goals helps to ensure that an appropriate priority is set and a new dynamic opens up between manager and employee.

Naming your program ambitions in detail also informs program design. Recognition takes place in day-to-day management. Unlike payroll or tax reporting, for example, a recognition system is not a matter of buying a new software package. Hard-to-measure factors such as the adaptability of your line managers can play a significant role in the effectiveness of recognition. Early in your design of a recognition program, ask the questions that set a size and scope for your efforts, such as:

* How aggressive do we want to be? (e.g., How soon do we hit goal A? How soon do we hit goal B?)
* What resources do we need to start and continue the program?
* How fast can our managers learn this new skill?
* What other programs are in the pipeline at this time?
* Which departments (e.g., training, communications, finance) need to be involved in ongoing operation of the program?
* How do we report progress?
* What is the best rollout path?

Answering these questions in detail early on means a more cost-effective and powerful program later.

Once you have clear goals, translate those ambitions into measurable achievements to make recognition meaningful to executive leadership. Recognition metrics should:

1. Reflect what's important in the corporate culture based on its values.
2. Measure significance relative to your strategic goals, such as changes in productivity, cost, and retention.
3. Show how certain recognition behaviors drive employee performance (which also identifies best practices within your culture).
4. Link recognition to the organization's financial statements and corporate goals.

Examples of strong strategic metrics include:

* More than 90 percent of employees touched by the program (as nominators or recipients)
* One-year survey confirms that 90 percent of employees agree that "the program helps motivate sustained high performance"
* Program reaches all geographic and demographic groups of the organization
* Award distribution matches performance bell curve (see Tactic 6 below)
* Award frequency met by managers within six months of training
* Six-month survey shows 90 percent of managers participating
* One unified system meets budgetary goals at six- and twelve-month milestones
* Two-year survey shows double-digit increase in employee satisfaction
* Two-year survey shows double-digit increase in employees qualified as engaged
* Company values selected as award reasons by division, region, or business line, as appropriate

Measurement means relevance. Without it, any project tends to justify itself. The lonely corners of companies are cluttered with once-promising initiatives that lack measured success or failure. Often in recognition, standards of success are applied later in the game, simply to legitimize the project, instead of at the outset; this makes the metrics irrelevant.

The bottom line is, if you don't know what you're working toward before you begin, how will you know when you've arrived?

Determine your metrics *before* execution begins, then faithfully report against those metrics on a regular schedule, even if the outcome isn't what you hoped. Negative results can be the most valuable, as they show you the areas where you most need to improve. In companies where such failure is permitted, continual improvement is possible, not just in a recognition program but across the board.

If your organization has project managers, use their tools (or better yet, get a project manager on the recognition team) to keep the setup and execution on schedule for a smooth program launch. Then use periodic surveys to measure such factors as:

* Employee and manager participation
* Number and/or percentage of awards given
* Size of awards given
* Program budget
* Impact on employee morale
* Impact on customer satisfaction
* Impact on productivity
* Follow-up actions taken (if the program reveals management or employee problems)
* Impact on employee attraction, retention, and turnover
* Impact on engagement

Identify recognition opportunities as you implement the program. Recognize managers who understand and embrace the program, who realize improvements in their staff as the program continues, and who encourage and teach recognition as mentors to other managers. In other words, make progress in recognition important enough to recognize and reward!

TACTIC 2:
INVOLVE PROGRAM PARTICIPANTS AND INVITE THEIR INPUT
A program that improves employee alignment and commitment needs input from a cross-section of employee leaders, from high-ranking executives to hourly workers. Individuals ultimately implement culture change, so gathering input from all levels early is helpful. The owner of the recognition program will do this informally or systematically (such as surveying a random sampling of employees) according to the organization's traditions.

As you clarify your vision for recognition, balance the need for broad input against the need for a process that moves forward. We've seen different clients implement this differently. For example, an advisory board made up of employees, managers, and executives can provide early suggestions that higher-ups might miss. Middle managers and line employees are especially aware of the invisible but powerful social architecture of your organization, which is a necessary component of cultural change.

The common theme in successful program design is that the program owners ultimately have authority to move ahead. If everyone you survey has a vote (or veto) you'll never get out of the visioning phase.

Today, social architecture has become more fluid and fast-moving, thanks to the rise of social media, greater employee mobility, and greater openness of information. This macro-organization of many efforts, all of them moving in the same direction, is happening today in your organization. Communication, feedback, reinforcement of opinions, and input on ideas—social architecture touches every employee and activity.

MYTHBUSTER

✳ ✳ ✳

It's commonly believed that having a broad sweep of employees in every decision is democratic and thus, most effective. However, in our experience, things like decisions about graphic design of program communications can get bogged down in endless debate. For example, if you poll employees about specific rewards, they will always reply, "cash," and expect it. We've described the many reasons cash is not the best recognition reward, but don't set yourself up to teach everyone that before you implement your recognition program. If the topic comes up in a smaller advisory board setting, you can educate that group.

Keep the conversation on the topic of values: Why do employees support the company's overall mission? Hear in their words what motivates them, and who is most engaged and effective. You might survey employees to establish benchmarks in employee sentiment. If your organization is large or diversified, enlist helpers across divisions. If your organization is global, it is essential to get recommendations from locally based employees on such matters as local culture and preferred style of recognition. These early participants can become advocates and mentors to the larger company as you roll out the program.

TACTIC 3:
START THE TEMPO AT THE TOP
Contrast these scenarios:

Scenario A:
Bob and Mary sit on the HR team. They're recognition program managers, and they're low on the chain of command. They have no business planning experience. They don't understand how you develop a strategy in a company. They're told to get a recognition program going. An incentive company asks them, "Well, what's your budget? We'll help you spend that budget." So the company encourages Bob and Mary to put all their money into exciting, colorful posters, entertaining training for managers, and merchandise. The incentive company representative gives Bob and Mary a catalog of reward items and encourages employees to select from what the incentive company has in its warehouse. When you talk to the company's C-level executives about recognition, they say, "Oh, yeah, I think we have a program. That's Bob and Mary."

Scenario B:
Bob and Mary present a recognition program at a meeting of all top executives, including their boss, the head of HR. This group constitutes a green-light committee for every significant corporate initiative. Bob and Mary present the costs and benefits of a recognition program in hard financial terms. They go on to describe intangible benefits of the program, again showing statistics for things like increased engage-

ment and point out that intangible benefits produce tangible results. They cite Towers Perrin's finding that a 15 percent improvement in employee engagement correlates with a two percent improvement in operating margin. [xxvii] Their summary slide shows the one- and three-year returns on investment for the recognition program.

Now when you ask the C-level executives about recognition, they say, "We have a recognition program, and Bob and Mary report monthly on progress toward their ROI goal, like everyone else."

Strategic recognition is Scenario B, and you must have executive support to pull it off. Strategic recognition earns support because it engages executives where they live—in the realms of competitive advantage, high performance, and profits. The action steps outlined below speak to those concerns and earn executive buy-in.

BUILD A BUSINESS CASE FOR RECOGNITION.

A business case requires true management methodology. This means modeling a process throughout the organization, where hundreds or thousands of recognition moments are recorded, tracked, and measured.

A business case shows the cost of the program and its expected benefits. Your organization will have a preferred form of business case, and we recommend using that template to the greatest extent possible. At least calculate the true monetary cost of the problems a recognition program helps to fix (low productivity, high turnover, disengagement) and the benefits it inspires (increased sales, greater quality, and improved customer service).

Here's a simple example of a business benefit that recognition provides—decreasing employee turnover. A rule of thumb in recruiting says that the cost of replacing an employee who leaves is at least 30 percent of that employee's annual salary (factoring in the monetary cost of recruiting, lost productivity, ramp-up time for the replacement employee, and lost opportunity). With a $150 million payroll, a 20 percent turnover rate costs the company $9 million per year. Decrease that turnover rate to 15 percent—a reasonable goal for a recognition program—and you've saved $2.25 million in the first year on turnover alone.

You can perform similar planning on top-line factors such as improving market penetration rates and quality, for example, if you can measure how quality increases lead to greater market penetration and thus, revenue.

BUDGET STRATEGICALLY.
Strategic recognition requires sufficient budgeting, and our work with leading corporations puts that figure at around one percent of payroll. Now, we can hear you say, "Whoa, we don't have one percent of payroll set aside. How are we going to do that?" Back to business plan basics: You will not achieve the program's goals if the program is not appropriately resourced.

One of the failings of old-style recognition is its implication that recognition programs are more or less the same. That's like saying, "Let's buy any payroll system because they're all alike." Recognition programs, like payroll systems or any other enabling technology, have to justify their cost based on the value they generate, which is the point of modeling and measuring.

Like any enabling technology, strategic recognition bears a flexible series of costs—more so than most, in fact, because the cost of awards is so broad. (We'll provide guidelines for this in the "Match Awards to Achievements" section below.)

It can be done incrementally. Going from zero to one percent in one year can be difficult, and so some companies start with a budget of 0.5 percent of payroll for their recognition program, and then increase by 0.1 or 0.15 percent per year. It can also become budget-neutral as part of a total rewards strategy. In a nonrecessionary year companies typically allocate three percent of the payroll budget for merit increases. So as part of a total rewards strategy, you might re-allocate 0.15 percent from the merit increase to the recognition program.

Incidentally, most large companies are already spending one percent of payroll in ad hoc recognition programs; you just don't know it. Those well-intentioned, informal acts of recognition—taking the department out for dinner, or buying tickets to the ball game as a "thank you"—tend to be obscured as expenses in travel and entertainment budgets. We've known CFOs who have tried to separate these expenses into a real view of what informal recognition costs, but they never get to the bottom of it.

DRIVE TOWARD ALIGNMENT AND COMMITMENT.
Strategic recognition will need alignment across all top executives, and commitment from them to see it though. It's one challenge to get a budget granted, and another to command the attention and time of busy leaders in implementing a recognition program. This might not seem logical—what's worth spending money on is worth supporting—but nonetheless it's possible to see recognition crowded out by other strategic programs. We believe executive participation elevates the program to strategic status. It increases general participation, focuses managers, and battles the unfortunate tendency to say "yes" without real commitment.

INCLUDE RELATED CENTERS OF EXPERTISE.
Resources like budgeting, cost containment, program management, marketing, and compliance with different global legal and tax regula-

tions should be sourced from within the company if the HR department doesn't have the expertise. Get this information in advance and build it into your recognition plan before you present it to executive management.

POSITION RECOGNITION AS A LONG-TERM PRACTICE.
A strategic culture of recognition never ends, so you should be prepared to discuss its long-term costs and benefits. Three to five years is sensible planning for a company-wide strategic recognition initiative.

You will need to identify stakeholders and leaders. Our best practice studies recommend the following configuration of roles and responsibilities:

Senior executive sponsor(s)—the global program "champions":
For greatest program impact, one or more executive sponsors must be accountable for delivering all the program's components in accordance with the vision. They must agree to monitor and discuss the quarterly executive dashboard and take action to guarantee the program moves forward like any product initiative. The company's head of HR at the very least should be an executive sponsor. Executive sponsors:
* Drive global program awareness at a senior level;
* Sign off on program design and implementation;
* Validate global program goals; and
* Support the program's implementation over time.

Finally, executive sponsors should have a personal presence in your communications. If you send out a monthly email listing who's been recognized that month, it should go out with the signature of an executive sponsor. A personal communication from an executive to an award winner has a huge impact on the winner and on the program's reputation.

In-house program manager(s): Whether this is Bob and Mary in HR or the senior director of operations, this person has ultimate accountability for designing the program and launching it to greatest effect. This is also the main contact with any outside vendors and the person who chooses what models to follow. Program managers:
* Define goals, objectives, criteria, processes, benchmarks, and measurements of success;
* Drive company-wide consistency and equity of rewards; and
* Drive the program globally (in a large firm).

Local or departmental program "champions": Program champions, as symbols of executive support, are critical for two reasons. First, if members of the executive team are divided on the importance of recognition, there's a distinct possibility of neglect or a benign brand of sabotage—the death of many initiatives. Second, local champions are the experts when it comes to resolving cultural differences, such as equity in award levels or traditions for recognizing and rewarding behavior. Champions provide strategic and tactical vision as well, which speeds the success of the program. Program champions:
* Position the global initiative locally (internal marketing);
* Drive program goals and objectives locally;
* Provide local validation of global program goals (big picture); and
* Mediate challenges resulting from proposed local changes to achieve global consistency.

In any configuration, executive support helps prevent the classic problem of "push vs. pull," that is, the tendency of executives to launch a program "push" with great fanfare by pushing messages, and promoting the program in meetings and through other venues. This is well and good, but the "pull" of seeing executives actually participating in recognition, going on the project teams, helping design the rewards, and walking the talk speaks far more strongly about the program's importance.

TACTIC 4:
ESTABLISH KEY INDICATORS OF SUCCESS EARLY
Business planning concludes with accountability. Whose budget gets spent, and who gets the benefits counted against his or her end-of-year review? Who will see to it that each component of the program is accomplished? Accountability is about doing what you say, standing by your planning, and being open about amending your plan as you go. It means quarterly reporting (at least that often) and openness about progress. It also means holding vendors accountable for the standards of your program.

Transparency helps maintain accountability. Describe to all employees how managers and even line staff will be held accountable for implementing the program. This goes beyond a supportive "let's all do our best" message. Publish the kickoff date well in advance of even having completed building the program. It's surprising the uplift you'll get from publicly stating that this program is important enough to the company that you're rolling it out like all other major initiatives, with a public and firm launch date. (Without a date, we find, companies go down the implementation path and they linger and they tweak and they worry and they debate. When the CEO has gone public with the date, the team gets motivated to hit that date and everything goes much more smoothly.)

Actual performance dashboards for managers create accountability. If your organization is using a dashboard system, it's wise to adapt that format to a recognition program. A dashboard should include the following data to ensure success:

THE STATED VISION

State the overall global vision for the program and describe, in metric/key performance indicator form, some far-reaching business deliverables such as "Employee engagement scores improve from current levels by 15 percent," "Employee satisfaction scores around recognition hit 85 percent," or "Q12 scores indicate X".

KEY PERFORMANCE INDICATORS

A dashboard report should be defined with the above company-wide strategic metrics at the top and the key performance indicators (KPIs) in graph form underneath. KPIs include overall penetration levels, penetration of each division/department, and operational necessities such as award approval/disapproval levels, enterprise budget targets, and similar data.

A VALUES DISTRIBUTION GRAPH

A values distribution graph illustrates the amount of awards that have been granted tied to specific behaviors for specific values in specific parts of the company. This graph and the data that it illustrates are the keys to making recognition strategic, because they quantify what heretofore has been hidden information about the company. The visual impact of this graph is especially helpful to managers who want to understand and cultivate culture.

This example shows the percentage of awards in three divisions of a company, focusing on the reinforcement of four key values. The meaning of the metrics is found in each department's goals and makeup. Note that Division 2 is receiving many awards for respect and integrity, but few for innovation. Division 2 happens to be the accounting group, so this award profile is excellent (if this were the product development group, we'd have quite a different story).

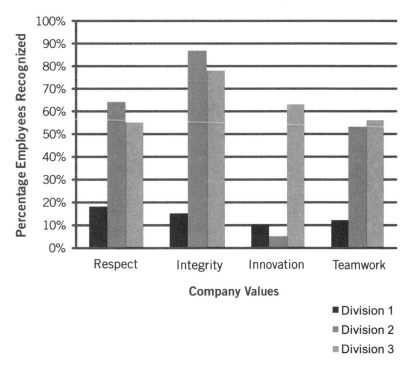

Values Distribution Graph

Division 3 is doing well across the board, with strong ties to each value. Division 1's performance is concerning. How is it that so little desired behavior is being recognized among all the company's most important values? Do the managers believe that 80 percent of the employees aren't aligned with the company's culture? Alerted by this graph, the division managers need to follow up and find the cause of the low numbers.

As we've said, strategic recognition uncovers which value-based behaviors are lacking and which are abundant. You might think of it as a brain scan of company culture.

It is essential that there is a strong correlation between recognition and the company values. If there exist some global, all-company business initiatives (e.g., Six Sigma or cost containment), then these can be added to the core values as award reasons. They can be separated into two bar charts monitoring the traction of award activity (linked to behavior) for these initiatives. This section (or an appendix to it) should suggest a range of award reasons and show the linkage between these reasons and the company values and global business initiatives.

The values distribution graph can function as an early warning system for company culture, in the same way that certain business data indicate turns in a national economy. For example, if a company's research and development department shows a low score on the value of innovation, something's going wrong—it could be that the managers don't understand the program, but it could also be an early warning that the spirit of innovation has waned. That early warning can cause management to correct the problem long before they'd otherwise be aware of it—typically when the business press runs stories stating, "XYZ Company's products are no longer innovative or interesting."

The early warning can be positive as well, such as finding a spirit of innovation in unexpected places, like customer service. Executives who discover hidden strengths through a values distribution graph can capitalize on those strengths. If your customer service group is innovative, you might join that capability with sales and marketing efforts to increase customer satisfaction and share of wallet.

The graph brings up another point that bears repeating: All this information is made more accurate, and more actionable, by high penetration of the recognition program. Capture the data from a program with 80 to 90 percent participation and you'll have a deep, dynamic view of how the company culture is driving results. An elitist recognition program, with 5 to 10 percent penetration, is statistically unreliable (if only the top innovators get recognized for innovation, how can you know if innovation is on everyone's mind?). High penetration must be closely monitored, managed, and incentivized.

The values distribution graph, in short, is an essential diagnostic tool to capitalize on the cultural insights produced by strategic recognition. Watch the graph, understand what it's telling you, and use it to take corrective action long before other indicators show you have a problem.

MYTH BUSTER

* * *

We understand that setting a target for the quantity of awards given each week, month, or quarter may be counterintuitive. Sometimes the response is, "Are we just giving everyone a prize? What's the value?" This fundamentally misunderstands the weight of awards in a system like this, and their place in the HR toolbox.

Remember, strategic recognition does not grant an award to a salesperson who hits his target at the end of the year (that's an incentive). Rather, it grants smaller awards throughout the year whenever he exhibits behavior that is aligned with the company values and objectives. Reinforcing those behaviors helps ensure that he will hit his target at the end of the year.

We've heard the objection that managers might give awards that are not deserved, just to hit their target. In our experience, nobody has ever given an excess of recognition. Even if this were to happen, the gap between recognition metrics and department performance will be quickly understood (an underperforming group that gets many, many recognition awards cries out for analysis—another way that metrics create accountability).

Don't underestimate the power of these metrics! Formally establishing a quota of quarterly awards ultimately reminds managers that recognition is fundamentally important in the company. Otherwise, they forget to reward those small, continuous, positive behaviors.

OPERATIONAL ACCOUNTABILITY

At an operational level, program metrics such as number of awards per department must be broken down into relevant targets for each division/department/country leader. These leaders will have their individual activity metrics communicated to them and will be accountable for monitoring their progress. Ideally this will be included in the MBO (management by objectives) plans for these managers. By setting penetration levels of recognition in their departments, reviewing recognition metrics regularly, and tweaking the program as necessary, the strategic recognition program will naturally become part of the management rhythm.

Other metrics track the progress of the program. Is it on budget?

Is your spread of awards as you predicted in number and value? Here are two examples:

Proper Budget Distribution for Meritocratic Recognition

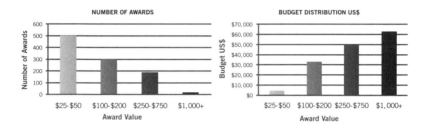

These two charts show the number of awards given on the left and the amount of budget invested in those awards on the right. The optimal charts should form a valley when viewed side by side as illustrated above.

The left side of this chart is descending, meaning that the program is designed to be meritocratic. The larger awards, which are far fewer in number than the lower value awards, gravitate toward the more impactful instances of employee behavior. This indicates larger awards are going to those employees on the high end of the performance bell curve.

The ascending right side of the chart indicates that the greatest numbers of awards are lower in cost, and the program is fundamentally designed to maximize visibility and penetration. Simply put, the company has opted to make the program more inclusive while still distributing the few high-value awards to the highest impact performers. If you get the balance right, you've maximized the program for penetration and visibility and you've maximized the program so that that top ten percent get most of the budget.

Imbalanced Budget Distribution
Creating an Elitist Program

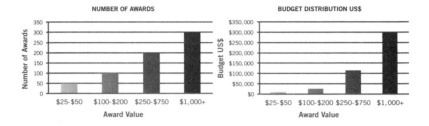

If these charts, viewed side-by-side, were to appear as above, then the distribution indicates an elitist program focused mostly on the highest performers. With the largest number of awards being given at the highest value, far fewer employees are recognized while the budget spent on recognition increases dramatically.

An executive can simply glance at these charts to assure they are trending in the right direction. If they aren't, the early warning system is ringing an alarm.

Some data will be purely operational, as in this example:

Budget vs. Spend

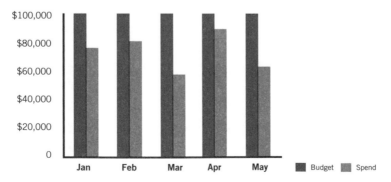

These data show that managers are keeping the recognition spending under budget—a good financial habit, as long as the other goals of the program (such as participation, distribution of awards) are being met.

Capture and interpret these metrics, and integrate them into your systems of performance management, compensation, and reward. (For that matter, recognize the excellent practitioners of recognition early, often, and publicly. That's really walking the talk of strategic recognition.)

These metrics are diagnostic and carry a powerful message that executive management cares about the program. Reports should go up the organization, ideally all the way to the CEO. This causes a cascade effect down the line; when everybody from the top down has a stake in hitting the goals, the clarity, urgency, and importance of the program magnifies. Everyone with authority is responsible. This cascade effect spreads learning through informal networks and through relationships with managers—the social architecture at work.

✳ The Don'ts of Strategic Recognition

The Ascent Group, a management consulting firm, ranks these qualities as characteristic of the *worst* recognition programs. Plan them *out* of your program! [xxviii]

* Inconsistency
* Untimely recognition / reward delivery
* Unclear program qualifications or criteria
* Perceived unattainable goals or uncontrollable goals

* Rewards don't match employee desires; limited choice or undervalued rewards
* Programs that don't include all employee classifications
* Programs with few winners
* Catalog award selection is limited

TACTIC 5:

PROMOTE IT OR PERISH

If strategic recognition is to become part of your company's DNA, it will require frequent communication. Andrew Liveris, president and CEO of Dow Chemical, put it this way: "At Dow, it is important that everyone understand why recognition is important, know how to recognize employees effectively, and make recognition a part of every day."

A part of every day. That means it has to be promoted, explained, advertised, and marketed to and by every employee. Fortunately, the nature of a recognition program makes it a natural fit for today's viral communications methods.

Recognition is an open-ended, continuous practice. Other HR practices like compensation, merit increases, benefits, enrollment, and performance appraisals tend to be annualized events, and employees have an acute, short-term incentive to pay attention and engage.

Recognition is more like a consumer product. Consumer marketing strategies include understanding your audience's current attitudes and beliefs, their desires and needs, and their willingness to learn. A long-held marketing belief states that a message has to be repeated many times, in many forms, for it to really sink in.

A well-designed program with a high penetration rate actually self-markets through the approval process. For example, I'm nominating you for an award. I get approval from my manager. When I give the award, my manager and the department manager are copied on the notice. Now four people are aware of the award. If five percent of the workforce receive an award every week, up to 15 or 20 percent of the workforce—and a high number of managers— are reminded of the recognition program every week. Awareness becomes self-perpetuating.

Moving the entire company from no awareness of the recognition program to high awareness requires a strong launch of the program and regular follow-up; as awareness becomes self-perpetuating, the formal message shifts to celebrating the awards themselves (as opposed to early communication, focused on educating employees about the system and its goals).

Launching strategic recognition calls for a *high-frequency, high-impact*, and *multifaceted* communications program. The goal is to increase program awareness and participation at all levels, because that leads to successful performance. Our experience shows that a successful launch happens when managers have seven to eight opportunities to see messages about the program in a short period of time, and in various media (online, in voicemails from the CEO, in email, in training, from peers, and in hands-on practice). We've also seen that 85 percent of successful recognition programs touch employees every two weeks in a variety of ways, including testimonials, success stories, training brush-ups, employee endorsements, and bright visual communication like posters and short video messages.

Engagement, excitement, and education are your goals, so use your organization's best communicators in a three-phase program:

PHASE 1:

PRE-LAUNCH—CREATE ANTICIPATION AND ENTHUSIASM

* "Tease" the program with personalized communication from the CEO and program leader to all employees, explaining the meaning and importance of the program.
* Distribute computer-based reminders about the launch of the program (such as screen savers, short messages on PDAs, a running electronic "zipper" or "ticker" display in the lobby, each describing a value to recognize).
* Set expectations with personalized voicemails from the CEO to division heads.
* Hold a call-in/web meeting with all division heads featuring top executives endorsing the program.
* Publish statements from managers testifying to their anticipation of the launch of the program.

PHASE 2:

LAUNCH—EDUCATE AND INSPIRE EARLY ADOPTION

* Send out a press release announcing the program.
* Send a mass voicemail to all employees on the day of launch.
* Send an animated email, linked to a video announcement made by a high executive.
* Publicize the program through Tweets, notes on LinkedIn pages, intranet notices.
* Release a video, on YouTube or via internal video channels, of executives describing how the recognition program affects everyone in the company.
* Announce contests such as an HTML email "nomination" campaign.
* Host a worldwide launch event.

On launch day, you'll get a quadruple benefit by giving a small award (of perhaps, $25 in value) to everyone. First, everybody in the company immediately knows about the program—and knows it's serious because you just spent money in the best way. Second, everyone experiences the pleasure of receiving an award; experience is much more memorable than, say, an email describing the program. Third, they get trained: They know what an award looks like. They have to redeem an award, so they have to go to the recognition website and go through the process of redeeming. And the fourth benefit is that you build morale because you've just given everybody in the company a gift.

PHASE 3:

POST-LAUNCH AND ONGOING—REMIND / INTEGRATE WITH DAILY ACTIVITIES

* Launch a company blog dedicated to celebrating recognition awards, including interviews with employees who have received awards!

* Send email or e-newsletter updates showcasing award winners and educating employees with tips about how they can be recognized.

* Hold informal chats about the program via social networks like Twitter and Yammer.

* Send press releases featuring interviews with "nomination challenge" winner(s)—those who have identified great employees to recognize.

* Hold regular company meetings that introduce managers who have used strategic recognition to improve performance.

* Conduct periodic reporting and showcase results of the program at three-, six- and twelve-month anniversaries, or keep a running tally of awards on the company intranet.

Branding

When recognition program leaders tell us they struggle to get their message out and unify employees behind their corporate mission, we ask first, "What do you call your program?" Sometimes the name is nothing more than a label: "XYZ Company's Recognition Program." At the other extreme, we've seen companies with dozens of different names for dozens of different recognition programs, all with names made up by managers, creating brand chaos.

We believe in a single, unique "brand name" for a company's strategic recognition program. Creating a unified brand promotes unity across divisions and borders. With one name known by everyone (allowing for translation), the company has a one-company voice and a one-company mission around which employees can rally. Branding your strategic recognition program in this kind of disciplined yet creative way is a chance to reinforce key values that make the company unique. As in consumer marketing, the brand ties multiple positive messages and emotional "hooks" into a single name, and strong branding helps make the program memorable for the long term.

Here are some more positive effects of a single brand:

* **Efficiency in marketing.** In a global company, multiple brands for the recognition program require duplicated effort, collateral materials, messaging, etc. We have encountered companies with as many as 20 separate recognition efforts with multiple names, which is confusing as well as wasteful.

* **Unity.** One brand across multiple business divisions promotes united effort and breaks down walls of geography and status. Everyone in every division who acts according to company values deserves recognition.

* **Culture reinforcement.** One brand, carefully chosen to speak about company values, reinforces the company's culture.

Our global clients that work in multiple languages wonder if a single brand can be translated effectively in all the languages used in their organization. There are three ways to do this. The company might choose one brand that can be translated into each language's idiom, such as "Thanks to You." Or, the company can use a common-language term—such as "Eureka!" or "Bravo!"—that is widely known (even in Asia). Finally, and less commonly, a company with worldwide operations might make up a name for its recognition program tied to the company name and that translates easily, such as "Globoforce+You."

Recognition program managers should plan to meet early on with the brand managers and the marketing department of their company to find a strong name for the program that conveys a message to wrap around the promise of your program.

TACTIC 6:

TOUCH AS MANY PEOPLE AS POSSIBLE, AS OFTEN AS POSSIBLE

Strategic recognition creates a big winners' circle. In old-school recognition programs, a few people received large awards. We've learned in practicing strategic recognition that you can achieve substantial improvements adding a low monetary award to your award portfolio. Giving many small awards in addition to a smaller portfolio of larger-denomination awards allows vastly higher penetration rates for your recognition program than large awards alone.

Companies have a bell curve of performance. There will always be an elite 10 percent of high-performing employees. If you recognize only that top 10 percent, then essentially the same group of people gets celebrated year after year. They receive big bonuses, take the trip to Hawaii, and so on.

Here's why that has less impact than you might think: That particular group of people already wakes up every morning very motivated to do their jobs. It's great to recognize them, but limit your recognition program to them and you create a very small winners' circle—one already inhabited by engaged, motivated employees.

Strategic recognition still rewards that top 10 percent (and as we have said before, with a properly designed program they will get the lion's share of the high-value awards). It also recognizes the 80 percent of people behind those top performers who carry the organization every day. These are the classic unsung heroes. They're the people who dutifully attend to their tasks, are consistently polite on the phone, and reliably complete the projects on time. Yet independent research shows that 65 percent of these employees have not received a recognition moment in the last year from their manager. A broad improvement in company performance requires bringing these people into the winners' circle.

Watson Wyatt Worldwide tells us in its 2008/2009 Work Report, "The key to driving productivity gains is increasing engagement among core contributors, who represent 60 percent of the typical workforce. Highly engaged employees are already working at or near their peak but are often limited by their less engaged coworkers. Focusing on engaging core contributors can improve both groups' productivity."

In his book *Winning*, legendary business leader Jack Welch, former CEO of General Electric, also recommends expanding the winners' circle: "The middle 70 percent are managed differently. This group of people is enormously valuable to any company; you simply cannot function without their skills, energy, and commitment. After all, they are the majority of your employees. But everyone in the middle 70 needs to be motivated, and made to feel as if they truly belong. You do not want to lose the vast majority of your middle 70—you want to improve them."

The bottom-line principle to affirm in a good recognition program is that everyone is eligible for recognition if they perform.

Frequent awards create an exciting sense of activity and raise awareness of the company's recognition program. The awards break the barrier of elitism among the rank-and-file, expand goodwill, and show respect. Frequent awards allow managers to make recognition a habit (and they get better at it). Furthermore, the bottom end of the bell curve—the worst-performing 10 percent of employees—learn what it takes to enter the winners' circle. If recognition is received by everyone who performs according to company values, those who truly don't care about performance lose the excuse that they are simply not members of a self-serving club.

When that middle 80 percent of employees are receiving recognition, democracy and meritocracy join to raise performance for the vast majority of employees. According to our case study produced with the Stanford School of Business, when 5 to 8 percent of employees receive some form of recognition every week, the program reaches that tipping point where most employees know the program and embrace its goals. In practice, some employees will receive two awards in a year, some will receive six. A large range of award values can be deployed, so participation isn't limited by someone's place on the organizational chart. The quality of service provided by customer service agents is as valued as the contributions of a division head when those actions demonstrate the company values in support of company goals.

As an example, one of our most forward-looking clients uses a mix of noncash awards redeemable in U.S. dollar-equivalent denominations of $25, $100, $500, and $1,000. This flexibility allows for frequent small awards and less frequent large awards. The average value of individual awards is approximately $100, with yearly total award spending of about $250 per employee.

Awards given to teams create a larger recognition moment, as more individuals participate while reinforcing the message that teamwork counts. When awarding a team, structure the award so that individuals receive equivalent awards and make the recognition moment a group event. (With a standard of living index controlling the monetary value of awards according to geography, cost of living, currency values, etc., the awards can be equitable across a global enterprise.)

Tracking a large number of recognition moments as part of a portfolio of awards gives relevance to the dashboards we've described because they render broader data across the company and along the bell curve of employees. The dashboard represents the effectiveness and the depth of the program in your little society that you call a company; the analysis of many awards, their causes and effects creates a nuanced and detailed look at employee engagement, productivity and morale.

A recognition program with frequent awards also creates immediacy; managers can create recognition moments quickly, tying the recognition and reward to behavior that's fresh in memory. We call this "high-speed recognition" because managers don't have to wait for the perfect moment or dither about whether the moment will be visible to all (the recognition moment can be repeated later, if needed).

Spontaneity adds impact and fun to a recognition program. Keeping the recognition moment simple ensures that everyone understands how it works. Combined with frequent communications, these many little, surprising moments create buzz around the program. It brings company values alive for the rank-and-file as a nonstop narrative of recognition illustrates the connection between company values and employee behavior. Soon, the program creates a thousand little stories of engaged employees being honored—what they did and, critically, why it mattered.

Which employee actions your managers recognize depends on the culture, values, and priorities of your company. Here is a list of examples:

* Helping someone
* Exhibiting behavior aligned to company values
* Going above job requirements
* High-performing teams (recognize everyone!)
* Passing a milestone
* Surpassing quality targets in a product or process
* Accountability (accepting responsibility and keeping promises)
* Customer service brilliance (measured by time to resolution, customer ratings, etc.)
* Bright ideas (even small ones)
* Making bright ideas a reality
* Promoting company values to the community
* Innovation (in products or ways of doing things)
* Initiative (not waiting to get started on a great idea)
* Leadership
* Inclusion and encouraging diversity of opinion, backgrounds
* Employee referrals
* Mentoring new employees
* Volunteering time to company-supported charities or good works

✳ ✳ ✳

Shorthand for managers:
If you would have said "Thank You!",
that's a recognition moment.

Remember the power that stories have to inspire managers and raise the profile of the recognition program. As your strategic recognition initiative launches, share examples of how company values are being recognized and rewarded. Circle back to your managers, asking them to expand a company-wide "knowledge base" of what actions they recognize, and why they are simultaneously important to employees and the bottom line.

TACTIC 7:
ENSURE A RECOGNITION MOMENT
What is a recognition moment?

Say "recognition moment," and our clients first imagine a ceremony where the entire department stands around and the manager gives out awards. Picturing this, and then thinking about achieving a high penetration rate in which five percent of the workforce is recognized every two weeks, our clients rightfully worry that there will be so many ceremonies nobody gets any work done!

Actually, the recognition moment is when an employee receives an award and, more importantly, *receives the personal message that accompanies the award.* That moment can happen publicly or privately, in person, via email, in a ceremony, or even at home via ordinary delivered mail. There are many options.

Three points are critical: The award is linked to a values-based activity, it's made quickly after the activity occurs, and somebody has taken time to write a personal message describing why the activity matters. The personal effort to write a message gives the award emotional impact.

Ceremonies delay the recognition moment and separate it from the activity that earned it, which reduces the impact of the award in addition to taking more time and sometimes becoming less personal. So where is there a place for ceremonies in a recognition program? Ceremonies are good for large team accomplishments, when many people are being rewarded. Little ceremonies might be scheduled into routine, for example when a manager reads the names and achievements of reward recipients at the end of a weekly team meeting. In a large company, an email or webcast might be broadcast periodically celebrating employee accomplishments large and small. (Again, this helps promote the recognition program and remind employees of company values.)

Be aware that public ceremonies aren't everyone's cup of tea. A recognition ceremony for an introverted person can be agonizing, which is more punishment than reward. Managers need to exercise judgment when choosing public versus private recognition.

MYTH BUSTER

✳ ✳ ✳

A public recognition ceremony can be counterproductive. Focus on the substance of the recognition—the personal message, the achievements being recognized, and the timing of the award—rather than the style of presentation.

A guiding slogan for effective recognition moments is *It's about the praise and not the prize.* It's about the manager conferring psychic income clearly, directly, and quickly. It's about strengthening relationships among team members and between managers and staff. Yes, the material award is important, but the recognition moment itself is primarily about psychic income—prestige, self-confidence, group cohesion, and pride.

Strategic recognition leverages your existing social architecture to help every employee perform at his or her best. Managers recognize and reward behavior, and the news of that action spreads. Suddenly, managers find themselves having a new conversation with employees about how their individual actions build financial and cultural value. Peers reinforce the message in conversation, email, and behavior. The social architecture is working to promote cultural values, and everyone profits.

The recognition moment is also a great moment of communication—positive feedback that can lead to a richer dialogue between manager and staff. Even if the moment comes in an email, the employee's response to the manager is a connection.

Not every manager is blessed with the warmth or personal presence to deliver the recognition message perfectly, but that hardly matters. We have found that there are several important behaviors to bring to the moment:

* Make the presentation culturally appropriate, especially in international divisions of global companies.
* Always tie the act being recognized to a specific company value, and to a long-term strategy.
* Convey the reason for recognition as a story. One classic format is known as SAR (you faced this *Situation*, took this *Action*, and got this *Result*).
* Tell how you feel about the employee, such as appreciative, approving, proud, gratified, glad to have him or her on your team.
* Check details before the moment occurs, such as name spellings and pronunciations, dates, manager and team members, and the division's big goals and/or targets.
* Celebrate winners and winning but do not create a "class system" of winners and losers. Everyone aware of the recognition must

believe they are eligible for the same reward.

* Focus on the inherent value of the employee's actions more than the material value of the award.

Recognition is a wonderful tool for making sure there is positive conversation between managers and employees. The implicit message: "I am recognizing you because you did something recently that contributed to the company in the following way. I want to make certain that you know that I and others notice, approve, and appreciate what you did."

TACTIC 8:
MATCH AWARDS TO ACHIEVEMENTS

In an organization where merit matters, awards vary to match the size of the achievement being recognized. People crave affirmation that their actions are important and meaningful, but understand that actions have different effects on the bottom line.

At the outset of a recognition program, managers are correctly concerned about deciding on the size of awards and ensuring equity among employees. While we've said that the praise matters more than the prize, managers still need guidelines. These can be set early in the design of the program, bearing in mind factors like budget, the differing value of awards across geographic regions ($100 buys more in Mumbai or Memphis than it buys in Manhattan), and the need to include all deserving employees.

Employees should have a simple concept of how different awards levels are set. One model: Award level A = extra contribution; Award level B = unusually strong contribution or unusually tough challenge met; Award level C = extraordinary contribution.

Here is a list of best practices for setting your award guidelines:

* Keep the number of award levels as simple as possible.

* Establish award levels that relate easily to the degree of the employee's contribution or achievement.
* Ensure that the award values are substantially different to simplify the awarding manager's choice.
* Provide corresponding examples of behavior for each award level for consistency of awards on a global basis.
* Keep in mind that the appropriate currency for a "compensation strategy" is cash, whereas the appropriate currency for a "recognition strategy" must be noncash.

To ensure equity and management of the recognition program without bogging down the program, do the following:

* Keep the number of approval levels to a minimum to ensure speedy award approval and to make sure the award is issued as close as possible to the time that the behavior was exhibited for maximum impact.
* If possible, set a range of awards that managers of different levels can make spontaneously without asking for permission. Set and monitor the total number of awards budgeted.
* All awards must be tracked and reported upon; therefore, approvals should not become a barrier to wide use of awards.
* Low-value and peer-to-peer awards should require no approvals, where possible.

Be alert to the pros and cons of approval processes. While it's logical for your organization to establish guidelines for approving the spending on awards, don't let an approval chain delay most awards because speedily recognizing behavior is very powerful. If the COO has to approve most awards, that will take time and the award will lose some of its impact; furthermore, the message such delays send (that the manager who decides on an award needs to ask permission) can be counterproductive to the whole program.

Simple approval levels are almost always best for the great majority of awards in a high-penetration recognition program. Requiring managers to get just one sign-off on their awards should be adequate in most cases, except for the highest-value awards.

MYTH BUSTER

✳ ✳ ✳

Maybe approvals are overrated. How about going maverick? What if managers were simply given a recognition budget and guidelines on award levels, and didn't need approvals? As long as the awards are tracked and analyzed, the controls are in place to have an effective recognition program. If a manager goes over budget, his or her manager will quickly recognize a problem and correct it.

A side benefit of this practice: Managers gain more autonomy and accountability for their main job, which is... managing people!

TACTIC 9:

CALL ALL MANAGERS TO TRAINING

When implementing a recognition program, employees will require training in how the program works. Telling employees about the mechanics of the program—what it is, where to log in to suggest or collect an award, how to use an online wizard to track awards, and so forth—is the easy part. In a global company, this might also include information about tax considerations or budgets, for example.

But showing managers how and when to create recognition moments requires a whole different kind of training. We've seen that this is even simpler than the mechanics but causes much more anxiety over doing it right. People believe that because recognition is powerful, it's as difficult as other management disciplines.

This is a bit of a red herring. People understand the fundamental act of recognizing achievement, and if your program is tracking recognition activity, the program owner will quickly see where recognition is happening and where it's not. If you launch a program and let managers know they have to hit a target number of awards given, their awareness of value-affirming behaviors grows.

Your managers haven't earned an MBA in recognition, but they're not children, either. Nor is their job to entertain employees. Not everyone needs high levels of emotional intelligence for recognition to work. Employees already know their managers' temperaments. As long as a manager's appreciation is sincere, fair, and deserved, employees will get the message. While training can help the less socially adept manager's use of recognition, most of your training should focus on managing strategic recognition as a management discipline systematically across divisions, locations, and cultures.

Summarize training by telling managers, "Catch your people doing the right thing." That said, there are some good ways to reinforce the essentials of strategic recognition:

* Teach mechanics, such as the hands-on operation of the recognition approval and reporting system, using every method available, including efficiencies like web-based interactive training.
* Train the trainers to understand the business implications of recognition and the differences between personal, enterprise, and strategic recognition.
* When choosing between speed and formality of recognition,

choose speed. Launch the program and watch your managers figure it out as they try to hit their awards targets.

* Include baseline recognition practices for all training of new managers.
* Have managers who are especially talented at recognition act as mentors to others and ambassadors of the recognition program as a way to maintain everyone's enthusiasm and improve their recognition skills.
* Collect and communicate real-world accounts of recognition moments throughout the company—train by example!

Continue the enthusiasm generated in training through your ongoing communications program. Remember that while training can instill new skills, managers will make recognition a daily practice only if they are convinced it brings results, so broadcast those results in emails, newsletters, and ongoing management training.

TACTIC 10:
OFFER A GREAT CHOICE OF REWARDS

Employees are individuals—this is a fundamental principle of strategic recognition. Each employee makes unique contributions, has unique tastes, and desires uniqueness. The best managers, and indeed the best global organizations, learn that rewards are most effective when they honor the individual.

Here is where vendors of old-fashioned recognition programs operate using a flawed business model. Their revenues derive chiefly from selling preselected merchandise, which cannot cater to so many unique individuals. This business model dilutes strategic recognition's core belief that it's about the praise and not the prize, because old-fashioned recognition is tied to the prize.

Even the largest merchandise-based programs are limited in their offerings to the range of products being selected by a purchasing agent working for an incentive company. That might be a thousand items, but compared to the range of choice consumers have now in shopping online, it's a limited catalog. It's not offering great value by comparison to, say, Macy's, where a professional buyer can find items that are current, fashionable, and focused on the local Macy's customer.

Merchandise-based programs were designed before the advent of the multicultural, global organization, and they make little allowance for differences among cultures. A crystal trophy might sit well on a desk in Paris, but does it have the same meaning in Toronto or Miami?

Is the prize (and even the praise it represents) delivering a strong and lasting impact for its cost? How do you know? How much cost is wasted by shipping merchandise around the world?

MYTH BUSTER

✳ ✳ ✳

Employers see colorful merchandise catalogs and think, "These items are specifically made for recognition, and therefore they are best to use for awards."

The game has changed since the heyday of award catalogs, however. First, the Internet makes it possible (and even probable) that employees will quickly determine the cash value of merchandise awards—and who's discounting that item this week. Second, removing the power of choice from employees is counterproductive and can produce negative sentiment, as if telling the employee she didn't have the sense to make a choice.

The Internet has created a world of choice for all consumers, including your employees. With literally millions of objects available at the click of a mouse, consumers have become extraordinarily brand- and cost-conscious, and they expect a wide range of choice in everything from music to clothes to restaurants to travel. Note that this includes experiences as well as items, and experiences create strong positive emotions for people. For many, a visit to a restaurant or spa can carry greater value than a new possession.

Strategic recognition embraces this trend and magnifies the moment of recognition with the power of choice. In a partnership of goodwill, the giver determines the value of the award, and the recipient determines the final object or experience awarded (typically through the use of merchant-specific gift cards). In this way, the recipient becomes a participant in his or her own recognition, not a passive-if-gratified spectator.

Imagine the conscious and unconscious thoughts of an employee who's just received such an award:

Employee A shipped a critical product early last quarter, and her manager recognized the accomplishment with a nice award. How will she use it? Since she worked many nights and weekends to ship the product early and was absent from home a lot, perhaps she'll choose a spa visit or weekend getaway for herself and her spouse.

Employee B travels constantly as part of his job, and he might want those new noise cancelling headphones he has seen in airports. (He'll really enjoy them whenever he's trying to work on a flight while a fussy baby shrieks three seats away).

Employee C receives an award and enjoys speculating on what to choose: Maybe he wants to get high-end football cleats for his child with a gift card to Zappos.com or GigaSports. Perhaps his daughter will need a dressy blazer for her new job—he can get that with a gift card to women's clothing retailer Ann Taylor. Every time he contemplates one of these choices, he subtly relives the meaning of the award—the recognition, appreciation, and connection to value-based performance.

Other advantages of offering such a broad choice of awards:

It makes sense financially.
More value goes directly to employee awards because cost of goods, shipping, etc., is minimal. You're spending less money on ordering and shipping merchandise all over the world.

It makes sense operationally.
Awards can be tracked using budgeting and management software.

It promotes the perception of fairness.
The monetary value of awards is transparent, which promotes feelings of equity among employees and helps managers keep things simple.

It uses the best of both cash and noncash awards.
Like cash, the award value is tangible and known, but it is also directed to discretionary spending and benefits (e.g., not used to pay the electric bill).

It employs both local and global control.
While global budgets can be determined and managed, cultural differences among regions and divisions magnify the need for local customization. Managers and employees enjoy wide discretion, and the CFO can sleep well at night.

* * *

✳ ✳ ✳

Conclusion

ENSURING STRATEGIC
RECOGNITION SUCCESS

We've seen strategic recognition work around the world because it is built around the realities of today's corporations and the mindset of today's employees. Almost any recognition program brings benefits, but a program that is planned, launched, and operated as a major management initiative brings competitive advantage in abundance. In a world where most HR tools have been commoditized, why would a business striving to compete do anything less?

As you prepare to design and launch a program of strategic recognition, review the following checklist of essential building blocks of a truly successful recognition program. Double-check that these five elements of your program are in place. When they are, you can give a green light to the program launch and bring truly strategic recognition to your organization.

☑ 1. THE STATED VISION

Confirm that you have stated the overall, global vision for the program and describe, in metric/KPI form, your program's targets. These can be something like "Improve employee satisfaction scores from current levels by 15 percent," "Increase employee satisfaction scores around recognition and appreciation by 30 percent," "Achieve employee satisfaction scores related to recognition of 85 percent," "Achieve Q12 scores of X," etc.

✔️ 2. EXECUTIVE DASHBOARD

Create an executive dashboard report that uses real data to monitor progress in meeting the targets established in the previous section. The report should list the company-wide "vision" metrics at the top and the nuts and bolts metrics in graph form underneath (these include overall penetration levels, penetration of each division/department, and operational necessities such as award approval/disapproval levels, enterprise budget targets, etc.). To take this one step further, draft an executive dashboard report with the data you anticipate six months after the recognition program has been launched. When you get to that date, compare the projection with the reality.

✔️ 3. VALUES DISTRIBUTION ANALYSIS

Generate values distribution analysis charts from the data you collect. Executive leaders agree to monitor it and take corrective action if the results identify any problems. As we've said, strategic recognition's value can lie in revealing which values are lacking as well as which are abundant. It is essential that there is a strong correlation between all awards and company values. If there are some global, all-company, business initiatives (e.g., Six Sigma), then these can be added to the core values as award reasons.

✔️ 4. EXECUTIVE SPONSOR

Make sure that there are one or more executive sponsors in place to ensure that the strategic recognition program delivers what it promises. The executive sponsors will need to agree to monitor and discuss the quarterly executive dashboard and take any corrective action that is needed. Of course, they cannot do this if they have not been empowered by top leadership to do so, and they will not be effective if they do not agree with the targets that have been established. Often the head of HR is one of the executive sponsors, but there should also be someone from another division of the company.

☑ 5. LEADER ACCOUNTABILITY

At an operational level, program metrics must be broken down into relevant targets for each division/department/country leader or grouping (depending on the company). These leaders could be the people who approve the awards, for example. Without targets, these leaders will not have true accountability for the success of the recognition program. These leaders will have their individual activity metrics communicated to them and will be accountable for monitoring progress. Ideally, this accountability for the recognition program will be included in the management by objectives plans for these managers. By setting targets for recognition in their departments, reviewing recognition metrics regularly, and tweaking the program as necessary, companies can make the strategic recognition program become a natural part of the management rhythm.

✳ ✳ ✳

✳ ✳ ✳
Afterword

THE PROMISE OF
STRATEGIC RECOGNITION

Recognition, appreciation, and rewards are fundamental to effective management and ought to be practiced strategically. For its first hundred years, however, the formal practice of recognition focused exclusively on the relationship between the manager and the employee. Enlightened human resource professionals have gone a little further in recent years, training managers to provide encouragement, and in the hierarchical cultures of business, helping managers balance carrots (rewards) and sticks (consequences, which typically meant withholding rewards) among their employees. This seemed an adequate practice when employers believed peak performance required only training, command, and compensation.

Those days are gone. Today's employee seeks self-determination. Lifelong employment is a thing of the past and, in response, the best talents in your company think of themselves as "temporary workers"—engaged not out of loyalty to the company but because their needs are being met for both tangible and psychic income. They want their work to have meaning, not just monetary reward.

Effective managers know this (sometimes unconsciously) and use individual recognition to teach and bond with the employees they must guide day-to-day.

Across the organization, enterprise recognition grew as "more of a good thing," and HR typically spread the good practices of recognition across the company. Relationships between managers and staff become richer and company goals were more widely discussed. Most important, recognition and appreciation began to be directly connected to company values and goals. Practiced across the entire company, even globally, enterprise recognition became more than "throwing a bone" to a good employee. Enterprise recognition flattened the moral hierarchy of the corporation by making values-based behavior everyone's job.

Enterprise recognition, while effective, is barely connected to the other management systems that govern the company. Strategic recognition, in which the "soft" behaviors of recognition—which worked well at motivating employees—finally merge with the "hard" management practices of measurement, goal-setting, analysis, and strategic execution.

The promise of strategic recognition is thus multilayered and multidimensional.

* For the employee, it teaches the connection between behavior and values; it increases morale and encourages engagement, loyalty, and attachment.

* For the manager, it compels thinking about which behaviors actually embody company values. This is new, for while it's easy to like certain behaviors, the manager must actually choose which employee actions deserve recognition. To manage is to choose. To manage is to get work done through others. It makes the manager more effective, more accountable, and more directed.

* For the executive, strategic recognition finally—finally!—makes a direct connection between encouraging certain values and knowing if and where those values are being lived by employees and

managers. It brings the long-mysterious gifts of "people skills" and the proven need for "hard data" together, ending the false but enduring conflict between the "humanist" and "realist" schools of management.

Strategic recognition promises to help companies gain competitive advantage, foster employee engagement, improve performance, and increase profits. In a business world growing increasingly interdependent, with organizations growing less command-driven all the time, it is the supple and powerful new way to enrich and manage culture. Strategic recognition is more than a technique; it is a mission.

✳ ✳ ✳

✱ ✱ ✱
Endnotes

i   *New York Times*, Sept. 5, 2009 (http://www.nytimes.com/2009/09/06/business/06corner. html?_r=1&pagewanted=2)

ii June 16, 2009 (http://blog.threestarleadership.com/2009/06/22/home-depot-at-30-a-lesson-in-corporate-culture.aspx)
June 22, 2009 (http://blog.threestarleadership.com/2009/06/16/lessons-from-the-rise-and-fall-of-delta-airlines.aspx

iii "Around the Globe the Desire for Meaningful Work Triumphs Over Pay, Promotion and Job Choices," Feb. 25, 2009 (http://www.kellyservices.us/eprise/main/web/us/hr_manager/document_center/kgwi_meaningfullwork_2_25_09.pdf)

iv *The Conference Board Review*, Fall 2009 (http://www.tcbreview.com/beyond-the-handshake.php)

v   *Winning*, by Jack Welch with Suzy Welch (HarperBusiness, 2005) (http://www.welchway.com/Principles/Differentiation.aspx)

vi "The Zappos Way of Managing," *Inc.* magazine, May 1, 2009 (http://www.inc.com/magazine/20090501/the-zappos-way-of-managing.html)

vii *Closing the Engagement Gap: How Great Companies Unlock Employee Potential for Superior Results*, Julie Gebauer and Don Lowman (Portfolio/Penguin USA, 2008)

viii Knowledge@Wharton, July 8, 2009 (http://knowledge.wharton.upenn.edu/article.cfm?articleid=2280)

ix "Making Sure the Shoe Fits at Zappos.com," *New York Times*, Nov. 6, 2008

x   This is primal stuff, right in our DNA. Recent studies have focused on more primitive but very real versions of these needs in other mammals, such as primates. See *The Wauchula Woods Accord: Toward a New Understanding of Animals* by Charles Siebert (Scribner, 2009)

xi See *The Seven Habits of Highly Effective People*, by Stephen R. Covey, pp. 190-203

xii *American Insights Workplace Survey*, Adecco Group North America, cited in "Corner Office, with a Side of Employee Disdain," *Human Resources Executive Magazine*, Nov. 9, 2009

xiii "Hating What You Do," *The Economist*, Oct. 6, 2009 (http://www.economist.com/businessfinance/displaystory.cfm?story_id=14586131)

xiv Ironically, great companies see HR excellence as critical to success, and whether HR is called a cost center or not, it has the same resources and expectations as other strategic departments.

xv "Making Talent a Strategic Priority," McKinsey & Company, 2007

xvi "Productivity and Motivation," *Personnel Today*, July 1, 2008: "The unanimous finding of our survey was that appreciative colleagues have a positive effect on productivity: Two-thirds believed they were a lot more productive when given encouragement by their workmates."

xvii "Executing Strategy: Alignment Makes the Difference," Towers Perrin International Survey Reseach, June 2009

xviii Define, Measure, Analyze, Improve, Control

xix "At Yum Brands, Rewards for Good Work," *New York Times*, July 11, 2009

xx "Recognize Results: Drive Success through Employee Recognition," study by OfficeTeam and the International Association of Administrative Professionals, April 2009

xxi "Saying Thank You Boosts Employee Engagement," study by the Institute of Leadership & Management, April 23, 2009

xxii "Compensation Programs' ROI Highlighted by Study," by Stephen Miller, Society of Human Resource Management, May 2006

xxiii "Human Potential Untangled," *T+D* magazine, April 2009

xxiv This is the subject of a famous essay he wrote several years ago, which should be required reading for every manager, especially in HR.

xxv "Prepare Now or Lose Your Best Employees Soon," study by Deloitte Consulting on *Forbes*.com, Aug. 12, 2009 (http://www.forbes.com/2009/08/12/talent-employees-retention-leadership-ceonetwork-deloitte.html)

xxvi "A Joint Survey by WorldatWork and National Association for Employee Recognition," September 2003

xxvii Towers Perrin's 2007–2008 Global Workforce study

xxviii "Reward and Recognition Program Profiles and Best Practices 2008," Ascent Group

* * *

✳ ✳ ✳
About the Authors

Eric Mosley

As co-founder and CEO of Globoforce, Eric has been directing the path of Globoforce as the innovator in the strategic employee recognition industry since the company's beginning. His vision to raise employee recognition from a tactical, unmeasured, and under-valued effort to a global strategic program with clear measures for performance and success is now being realized in some of the world's largest and most complex organizations. Eric continues to cast the vision of innovation for the company and the industry.

As a recognized industry leader, Eric has personally advised some of the largest and most admired companies in the world. His work has been published in such varied publications as *Chief Executive Officer, Forbes*, and *Fortune*, and he has presented at industry and investment conferences in Paris, Berlin, Chicago, New York and beyond.

Eric brings a wide range of management and technology experience to his role as CEO of Globoforce. Prior to joining Globoforce, he established himself as an accomplished Internet consultant and architect having held varied management and technology roles in CSK Software, Bull Cara Group and Logica Aldiscon. He holds a bachelor's degree in Electronics, Computers and Telecommunications Engineering from the University of Dublin, Trinity College.

148

Derek Irvine

As Globoforce's Head of Strategy and Global Marketing Teams, Derek is a seasoned, internationally minded management professional with over 20 years of experience working across a diverse range of industries. During his career he has lived in many countries including Spain, France, Ireland, Canada, Sweden, UK and the USA. In his role as a thought leader at Globoforce, Derek helps clients set a higher ambition for global strategic employee recognition, leading consultative workshops and strategy setting meetings with such global organizations as Avnet, P&G, Dow Chemical, Intel, Intuit, KPMG and Thomson Reuters. An authority on the topics of employee engagement and recognition, he has been a guest speaker at worldwide industry and professional group conferences and is frequently published in media including *Businessweek, Workspan,* and *HR Management.*

While previously working with top-five management consultancy firm PA Consulting Group, he advised clients in corporate strategy, organizational behavior, marketing and corporate communications. Derek has also worked in consumer marketing for many years, working on the world-class brands of Johnson & Johnson and Jameson Irish Whiskey. Derek holds a Bachelor's degree in Commerce and a Master of International Business Studies degree from University College Dublin.

✳ ✳ ✳

✳ ✳ ✳
About Globoforce

Eric Mosley co-founded Globoforce in 1999 with the goal of reinventing the employee recognition industry for the 21st century multicultural, multi-generational global organization. Advocating highly compliant, easily governable consolidated recognition solutions that focus first and foremost on positive reinforcement of employee behaviors through company values, Globoforce has become the enabling technology to create a culture of recognition at some of the world's largest and most respected organizations.

By changing the way company leaders think about recognition from a "nice to have" to a critical business methodology for driving productivity, loyalty and profits, Globoforce is the acknowledged game changer in the industry. A key component of the company's success lies in its realization that the wants and desires of every generation, every culture, every employee is unique and highly personal. With the industry's largest selection of local languages and local, street-level reward options, Globoforce helps companies overcome geographic and cultural barriers to motivate their workforces and engage employees more fully in the mission, vision and values of the organization itself.

More information on Globoforce, its philosophy of recognition, and its unique approach to the employee recognition industry is available on the company's website at www.globoforce.com.